HOME
NETWORK
MANUAL

The Complete Guide to **Setting Up**, **Upgrading**, and **Securing** Your Home Network

MARLON BUCHANAN

HomeTechHacker.com

ISBN: 978-1-7355430-6-2 (paperback)
ISBN: 978-1-7355430-5-5 (ebook)

Edited by Graham Southorn

DEDICATION & ACKNOWLEDGEMENTS

This book is dedicated to my dad, Leon, because I miss him. I can picture my dad looking at me with a proud smile as I show him this published book. My dad would always tell me I can do it. Dad, I did it! I love you and I miss you.

I'd also like to dedicate this book to my mom, whose love of books, words, and writing inspired me to become an author. Thank you, mom. I love you.

TABLE OF CONTENTS

PREFACE

Growing up as a child of the 80s, I had no idea or concept of a home network. No one I knew had a computer at home, and I didn't even know what the Internet was. My first exposure to computer networks was in college where dozens of computer labs were spread across campus. We were able to send instant messages to other students, collaborate on shared files, and even play video games with people across the world. These capabilities blew my mind.

I saw the future, and there was no going back. For the first few years out of college, I never had more than one computer. I focused on making sure I could access the Internet from home to download music, send emails, and surf the web. I went from a dial-up Internet connection to ISDN to DSL, and finally to cable. My 1 Mbps download speed seems paltry by today's standards, but it was screaming fast to me back then.

As the years flew by, the dot-com boom (and bust) of the early 2000s changed the Internet and the world of home networking. My life changed too. I got married, and immediately there was a need for two computers at home to connect to the Internet. Back then, I just had a modem from my cable company plugged directly into my computer. I didn't even have the equipment to get two computers in my house onto the Internet. This is when I started learning about home networks.

I figured out that I needed a router to share my Internet connection. That was just the beginning. As time rolled on, we wanted to share files, add more computers, stream video and audio to devices, put computers and network devices in different rooms, add wireless devices to our network, and… the list goes on and on. The more I began to rely on the Internet the more I realized I needed to worry about security too!

Fast forward to today, and robust home and wireless networks are a requirement for modern living. Video conferencing, watching TV, gaming, and so much more—all rely on your home network. If you are the person who is responsible for your home network, then you know that members of your household take it for granted when the network works well and lose their minds when it doesn't.

Do you want your network to support your household and all the devices in your home? Do you want your Wi-Fi to be rock solid without dead spots or disconnections? Do you want to make sure all the devices in your home are protected from malware, ransomware, and other hacking? If so, keep reading. This book gives you all the tools you need to have the high-performing, secure, and reliable home network that today's technologies demand.

ABOUT THIS BOOK

Who Is This Book for?

Routers, modems, switches, firewalls, hubs. ADSL, fiber, cable, cellular. The terms bandied about in the home networking world can become overwhelming and confusing in a hurry. Even if you are familiar with most, or all of these terms, combining them to get the best network possible can become overwhelming. In today's world, being able to set up a home network is more important than being able to set up a TV. Whether you live in a house or an apartment, are a boomer or a millennial, you rely heavily on your home network and the Internet for your quality of life.

This book is for anyone who wants to set up or improve their home network so that everyone in their home and every device can rely on it without constant fiddling. You don't want your streaming videos to buffer, and you don't want to be kicked out of your game because your Wi-Fi dropped. You want Zoom sessions with family and friends to be crisp and clear and your home network, computers, phones, and streaming sticks safe from hackers.

Perhaps you've already got your home network set up, but you are struggling in a couple of areas. Maybe you are looking to improve your home network but don't know where to start. Or you could be moving to a new home and starting from scratch. If any of these situations apply to you, then this book is for you.

When you are done reading this book, you'll know the following:

- What components make up a home network, and which components you need in your home
- How to make your home network fast
- How to have solid and reliable Wi-Fi throughout your home
- How to secure your home network from hackers
- How to evaluate your current home network and make improvements
- How to set up a home network from scratch and do it the right way

While this book is targeted at those who will build and maintain their own home networks, it's also useful for those who will have someone else build or maintain their home networks. Having the knowledge of how things work can help you better describe to your network technician what you want to accomplish and what problems you are having. Knowledge of home networking is important for everyone, even if you aren't building your own network.

How to Use This Book

This book steps you through building a home network from scratch. It starts by providing a foundation of knowledge and then using that knowledge steps you through the process of building and improving your home network. If you read this book sequentially, you'll learn the key components and technologies that make up a home network, and the way to first plan your own home network and then to build, secure, maintain, and improve it.

Many people reading this book won't be starting their home network from scratch. For those readers, this book can be used as a reference. The layout and chapters make it easy to quickly find the home networking information you're looking for.

Be sure to take advantage of the glossary, checklists, diagrams, sample network builds, and additional resources in the appendix. Home network technology involves a lot of jargon, and the glossary is there to explain all these technical terms. The checklists give you a quick way to track your progress in accomplishing the recommendations you'll find in each chapter. Finally, the additional resources in the appendix are links to up-to-date home technology resources.

Conventions Used in This Book

Here are the conventions used in this book, which highlight important information:

- **What I Do** – In several sections of this book, I give specific information and address the discussed topics in the context of my own home. For example, in the home network security section, I discuss some of the security steps I've taken. I'm not suggesting what I do is the best thing for everyone. Instead, these sections are intended to give you real-world, practical examples of what's possible.

- **Key Takeaways** – At the end of each part, I provide a checklist of the key information that you can use as a summary for quick reference.

- **How Do I Do That?** – Throughout the book, you'll find advice and solutions to common home network scenarios you are likely to encounter and decisions you may face.

PART 1

INTRODUCTION TO HOME NETWORKS

What Is a Home Network?

A computer network is a group of computers that share information and resources with each other. In many ways, a computer network is just like your network of friends and family. You share stories and information with your friends. Sometimes you ask your friends and families to tell you things, like what movie they recommend. Sometimes you tell a friend something, and that friend tells another one of your friends the same thing. Computer networks work the same way. Sometimes a computer sends information to another computer in its network. Sometimes it receives information from a computer in its network. Sometimes the information it sends and receives travels from computer to computer before reaching its final destination, just like your friend telling another person something you told them.

Using this friends-and-family analogy, you can view your home network like the network composed of friends and family that live with you. Some information you may want to share with the people in your home only (private family information, information that is useful for household members only, etc.). Some things you don't mind sharing or learning from your network outside your home (updates from extended family, discussing vacation plans, etc.). In your home computer network, there are things you may want accessible on your home network only (financial records, home videos, printers), and there are things you want to access and share with computers outside your home network (e.g., online games, websites, home videos you want to share with others).

A home network usually starts with your Internet connection, which is shared with all of the computers (this includes mobile phones, printers, smart devices, etc.) in your home network by your router. The router also allows all of the computers in your home to talk to each other. You may have more wired devices than the router has connections for. In this case, you probably have a network switch that connects multiple wired computers to your home network.

I know I just threw a lot of terms at you. Don't worry if you aren't familiar with all of them. I'll explain them in detail shortly.

Why Is a Home Network Important?

For most families, a home network is as important as having running water or electrical power in a home. Your home network is the primary way that all of your electronic devices connect to the Internet. The future is leading us to having more and more devices dependent on our home network. You are reading this book, probably because you recognize the importance of a good home network.

If absence makes the heart grow fonder, one way to think about the importance of your home network is to think about the things you can't do without it. Imagine not being able to make Zoom calls with family, friends, and work colleagues. Or not being able to play online video games with your favorite gaming console. Or not being able to stream Netflix to your TV! Without a home network, you can't do any of the following:

- Share an Internet connection with devices in your home
- Print to a printer in your home from multiple computers
- Stream video to your Fire TV, Roku, or similar streaming device
- Set up a Wi-Fi network, which means you can't fully utilize Wi-Fi-dependent devices (video cameras, smart speakers, tablets, etc.)
- Share files across computers
- Back up your important data to the cloud
- Work from home with a stable Internet connection
- Use most smart home devices (smart switches, smart plugs, smart lights, etc.)

This is the short list of things you can't do. These are the things many people have gotten used to doing every day. Imagine what life would be like if the only devices in your home that could connect to the Internet were your mobile phones!

Key Components of a Home Network

Before we get into the details and steps to build a fast, relia-ble, and secure home network, it's important to have a solid understanding of the key components. In this section, we'll walk through the key devices and network connection types that comprise a home network and discuss what they are, what they do, and why you would need them.

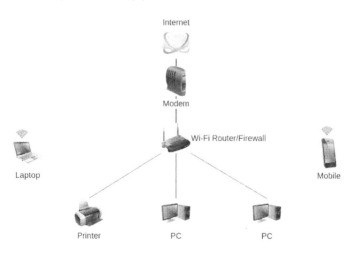

Basic home network diagram

Network Devices

Modem: Your connection to the Internet

Your home network's connection to the Internet is supplied by a modem. You usually rent your modem from your Internet Service Provider (ISP), but you can often save money by buying your own. It takes a signal from your ISP's network (usually a phone line, copper coaxial line, or fiber) and turns it into a standard type of connection, called an Ethernet connection, that most computers and routers support.

When ISPs list the "speed" of their connections, they refer to bandwidth. Bandwidth, in computer networking, refers to the overall data capacity of a connection. A 100 Mbps connection can transfer 100 megabits of information at a time in a second. Speed and bandwidth are used interchangeably in this book.

Router: Your most important home networking device

At its core, a router is a device that takes your Internet connection and makes it available to all the devices on your home network. However, today's routers do much more than that. They hand out and manage the Internet Protocol (IP) addresses of all the devices on your home network so that those devices can talk to each other. They act as the first line of defense from hackers getting into your home net-work, by providing a firewall. Most routers also provide a Wi-Fi network for your home.

Modern routers usually have a Wi-Fi antenna, a Wide Area Network (WAN) port that your modem plugs into to supply your router with the Internet connection, and a set of Ether-net ports to provide wired Local Area Network (LAN) con-nections.

Your ISP may supply you with a modem that also functions as a router. I typically advise people to avoid these devices or use them in modem mode only and buy their own router. Routers supplied by ISPs are often limited in functionality and are more likely to have security vulnerabilities—that is, they may not be as secure as you'd like them to be.

Your router has an enormous impact on the security, speed, and capabilities of your network. Choosing your router is one of the most important home network decisions you'll make when building your home network.

Firewalls: The key to keeping your home network safe

Network firewalls control what traffic can enter and leave your home network. For example, firewalls can block some-one elsewhere from logging in to your network and gaining access to all the computers on your network. Many can be set up to block computers known to be used by hackers from connecting to your network. Most routers come with a network firewall.

Personal computers, laptops, and even mobile devices can have their own firewalls in addition to the network firewall. These are a good second line of defense in case a hacker penetrates your network firewall or somehow gains direct access to your network (e.g., connects to your Wi-Fi).

There are many levels of sophistication to network firewalls, which we'll get into in the security section of this book.

Network switches: Adding wired connections to your LAN

Network switches, commonly referred to in networking terminology as simply *switches*, are devices that have multiple Ethernet ports that allow you to add wired devices to your LAN. You usually use one of the Ethernet ports to connect to your router, and the other Ethernet ports to connect computers to your LAN. Consumer switches usually come with 4, 8, 12, or 24 ports. They are called switches because they "switch" traffic coming and going from one port to another port.

Most switches used in home networks are *unmanaged*, meaning they are plug and play and there is nothing for the user to configure. They just work. There are also *managed* switches. Managed switches have some additional capabilities:

- The ability to prioritize different types of traffic (like video calls and gaming). This is usually referred to as Quality of Service (QoS).
- The ability to link multiple ports together to act as one port (link aggregation). This can help increase network bandwidth and provide a backup network connection (in case a network port fails) to a computer that also supports link aggregation.
- Support for virtual LANs (VLANs), which allows you to segment your network and control what traffic can go to different parts of your network. You'll learn more about the uses of network segmentation later in this book.

These are just a few examples. Managed switches come with more capabilities than these. They also come with a higher cost than unmanaged switches.

There are also Power over Ethernet (PoE) switches. These switches can send network traffic as well as power through the Ethernet cables to compatible devices. This is useful for powering devices that you install in locations that don't have an easily reachable outlet. It also simplifies cord management. Note that not all PoE devices are compatible with every PoE switch. Make sure to check whether your PoE switch and PoE device can work together. Generally speaking, look for PoE switches and devices that support the 802.3af/at standard.

Some people use the terms hub and switch interchangeably, but technically they are not the same thing. A hub works by blindly sending all network traffic to all computers connected to it and leaves it to each computer to decide which network traffic it should listen to. A switch examines the network traffic and sends only the traffic meant for a computer connected to a specific port, making it much more efficient. You'll have a hard time finding a real hub these days, so when most people say, "hub," they are probably referring to a switch.

One thing to note is that most routers contain a switch. The ports on the back of a router that allow for wired LAN connections are nothing more than an integrated switch.

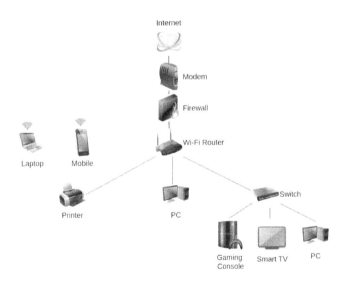

A network with a modem, firewall, Wi-Fi router, switch, and wired and wireless devices

Wireless access point: Adding wireless connections to your LAN

Wireless access points (WAPs) extend your LAN by providing a Wi-Fi connection point to your home network. They have wireless radios that broadcast a Wi-Fi signal for your wireless devices. They also have wired LAN connections that allow them to connect to your router and other devices using a cable.

You are probably thinking that's what a router does, and, in practical terms, you are right. That is because most routers also have an integrated WAP. However, not all routers provide a Wi-Fi connection, and in that case, you would want to buy a separate WAP to add Wi-Fi to your network. Another reason to purchase a WAP is to extend your Wi-Fi network to a place your router's Wi-Fi signal doesn't reach or to split the load of having many wireless devices on your network.

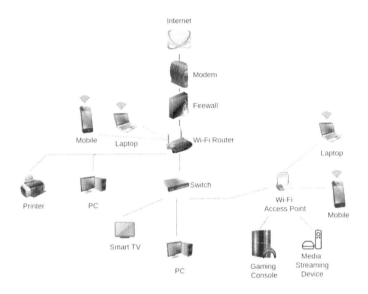

A network with a Wi-Fi access point. Dotted lines indicate a wireless connection.

Mesh Wi-Fi system: An easy way to have good Wi-Fi throughout your home

A Mesh Wi-Fi system is very similar to adding Wi-Fi access points to your LAN. One of the differences is that mesh systems usually come in bundles of access points that are specifically designed to work together. A second difference is that most mesh systems also want to serve as your router (although there are exceptions) whereas a wireless access point works with your existing router. A third and important difference is that mesh systems use completely wireless access points. You can place these satellite access points around your home to extend your Wi-Fi connection. Unlike a wireless access point, mesh satellites don't need a wired connection back to the router (called a backhaul). This makes them more flexible but sometimes a little slower than wireless access points. And there is a limit to how far away from the wired base unit the mesh satellites can be. A mesh system can provide a wired backhaul; however, when I reference mesh systems in this book, I refer to wireless backhaul systems.

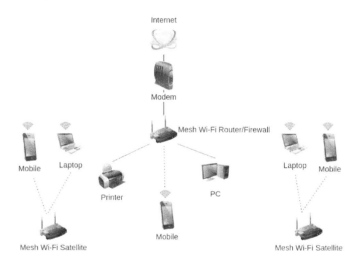

An example of a network with a mesh Wi-Fi router and satellites. The dotted lines indicate a wireless connection.

Wi-Fi extender: Extend your Wi-Fi signal wirelessly

A Wi-Fi extender, sometimes referred to as a Wi-Fi repeater, increases the range of your Wi-Fi signal. Like a WAP, it should be used to strengthen and extend your Wi-Fi signal to places in your home that don't have a signal, or where the signal is very weak. Unlike a WAP, it does not need a wired connection to extend your signal. The upside of this is that it adds some flexibility in where it can be placed. The downside is that it's typically going to be slower than a Wi-Fi access point, because it has to both send and receive (repeat) the data it receives wirelessly, reducing the overall throughput of the Wi-Fi network. By contrast, a WAP transmits the data it sends and receives wirelessly back to the network over a wired connection. Whether or not you'll notice the Wi-Fi speed loss depends on the efficiency of your Wi-Fi extender and how fast you want your network to be.

Network card: How your computer connects to your home network

An often-overlooked component of your home network is your computer's network card. Your desktop computer will usually come with a network card that allows for an Ethernet connection to your network. This card controls the speed at which your computer connects to the network. Most new computers will come with cards that allow for a gigabit (1 Gbps) connection to your home network. This is plenty fast for most of today's technologies.

Laptops, mobile phones, and tablets typically come with a Wi-Fi network card. This allows them to connect to networks wirelessly. Desktop computers can have Wi-Fi cards, and it is even possible to add Ethernet (wired) connectivity to mobile devices. But who wants a mobile device that is tethered? It may make sense for laptops sometimes. You're unlikely to want to connect a wire to your tablet or phone. We will discuss wireless connections and protocols in detail later in this book.

Network Connection Types

Ethernet cables: The most reliable connection

Ethernet cables make up the backbone of the modern home network. When you want the most reliable connection to your network and to the Internet, it is best to use Ethernet cables. They aren't subject to the environmental interference that Wi-Fi is. And, although Wi-Fi speeds are improving, Ethernet connections provide the fastest consistent speeds. Also, as I wrote earlier, Ethernet can be used to supply power (PoE) to devices.

Not all Ethernet cables are the same. They vary in quality and speed. When we discuss building your home network, we will cover these differences and which cables you should be using.

Wi-Fi: Connect to your home network anywhere in your home

Wi-Fi networks allow for connections to networks, using radio waves instead of wires. This means you can use de-

vices on your network without being near a wired connection. As mentioned earlier, many mobile devices don't come
with an Ethernet connection and expect to connect over
Wi-Fi.

Wi-Fi greatly increases the flexibility of connecting to your
home network, but it has drawbacks. Lots of environmental
factors (including microwaves, baby monitors, and neigh-
boring Wi-Fi networks) can greatly inhibit the reliability
and speed of your Wi-Fi network. Older Wi-Fi protocols are
significantly slower than newer protocols, and not all mobile devices support the faster protocols. Wi-Fi can be tricky
to get right, and we'll walk you through how.

Powerline: Use the wires you already have in your home to expand your network

Powerline networking uses your home's electrical infrastructure to transmit network signals. Powerline adapters
plug into electrical outlets in your home. Computers,
switches, and/or wireless access points connect to them via
Ethernet. You typically need at least two powerline adap-
tors: one that is near a network switch already connected to
your home network or that is near your router, and another
that is plugged into an outlet that is near the place where
you plan to extend your home network. Today's powerline
network adapters can get you speeds close to the faster Wi-
Fi speeds available today. Powerline networking works better in some homes and some outlets than it does in others,
so it sometimes requires a little trial and error to get right.

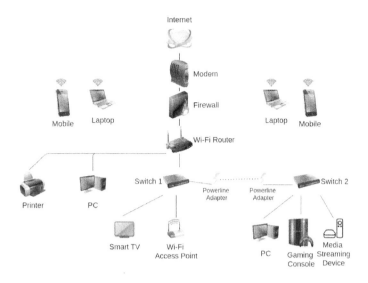

An example of a home network using a powerline adapter

MoCA: The most overlooked network connection

MoCA, short for Multimedia over Coax Alliance, is the name for a technology that runs network connections over coaxial wires. Coaxial wires are commonly used for delivering service to cable and satellite TV systems. Since most homes in the U.S., even older homes, have coaxial connections running through the home, MoCA presents an excellent opportunity to easily run reliable wired network connections. Even though the technology is still improving, MoCA connections can already achieve close to gigabit speeds, which is faster than the speeds available in most home networks.

You'll need to purchase MoCA network adapters to convert the coaxial lines to an Ethernet connection. Installation is as simple as connecting up your cable box. However, it's common for some of the coaxial lines in your house to be disconnected from other wiring (this is part of what the cable person checks and configures when they do an install). They usually all terminate at the same spot, and you may need some splitters or connectors to connect the lines you need. Also, many cable and satellite systems already use MoCA to provide network and Internet access to their set top boxes. In this case, you may have trouble running a second MoCA network across the same lines. You may be able to connect their MoCA network to your LAN and still use it. I used to do this when I had DirecTV.

Internet Connection Types

Dialup: The last resort of Internet connection types

When home Internet connections were first maturing, dialup was the primary way in which people connected to the Internet. Dialup connections require your computer to make a phone call to your ISP's server. You'll probably top out at a speed of .056 Mbps, which will make even simple web pages load slowly. Some extremely rural areas in the United States still have only dialup Internet access. If you have access to anything else, you should avoid dialup.

ADSL: Like dialup, but much faster

Like dialup, Asymmetric Digital Subscriber Line (ADSL) uses your home's phone lines to connect to the Internet. Unlike dialup, you can make and receive phone calls while you are connected to the Internet! You can also achieve respectable speeds (up to around 25 Mbps download and 5 Mbps upload) for your home. ADSL is a mature technology, but it tends to be less reliable than newer Internet connection types.

Cable: It's not just for TV

Cable Internet uses cable TV lines to provide high-speed Internet access to your home. It is one of the more popular U.S. Internet connections today because of all the investment in cable TV infrastructure over the last few decades. Modern cable Internet connections can achieve gigabit (1 Gbps) download speeds, but usually have much slower upload speeds, topping out at around 35 Mbps.

Fiber: The gold standard of Internet

Fiber Internet transmits light signals across fiber-optic cables to send data to and from your home network. Fiber is considered the gold standard of home Internet today because of its reliability, fast symmetrical bandwidth (often reaching 1 Gbps download *and* upload speeds), and low latency (how much time it takes for a signal to travel to its destination and back). Gamers really care about latency.

Not all fiber connections are equivalent. Fiber to the home, or FTTH, is the holy grail of fiber connections. This means that the fiber Internet connection goes straight into your home. Fiber to the curb (FTTC) means that your fiber connection goes to the nearest utility box (often at the "curb" of your home). After that, coaxial cables are usually used to send the signals to and from your home to the "curb." This introduces some speed bottlenecks. Finally, there is fiber to the node (or neighborhood, FTTN), which provides a fiber connection to multiple customers within approximately a mile or so of the node. Then, it usually uses ADSL to go from the node to your home. The further you are from the node, the slower your Internet will be due to limitations in ADSL technology.

Cellular: The future of Internet service

Cellular connections are best known for their use with mobile phones. Speeds from 4G approach ADSL and lower-end cable speeds and can be used for home Internet. Cellular companies usually cap how much data you can use in a month with most cellular plans (even if they call it "unlimited"). Some offer 4G plans meant specifically for home Internet, and they aren't subject to data caps.

With the advent of 5G technology, cellular Internet home service is poised to grow. Theoretically, 5G can surpass gigabit speed, although few, if any, reach that speed in practice today. However, they do rival faster cable download speeds (and surpass them in upload speeds).

Satellite: When wired and cellular connections aren't available

Satellite is another popular Internet technology for those living in rural areas that don't have access to ADSL, cable, fiber, or cellular connections. Satellite Internet technology involves three satellite dishes: one at your home, one in space, and one at the ISP's network hub. Your Internet signal bounces from your home to the ISP via a satellite in space. Sounds cool, huh? It has a few downsides. Speeds are often slower than cable (but not always), and weather and other obstructions can severely affect the reliability and speed of your Internet connection.

Important Home Network Protocols

A network protocol is an established set of rules that determine how data is transmitted between different devices in a computer network. Network protocols allow computers all over the world to communicate with each other, even if they are running different programs, using different hardware, and/or running different operating systems. Computer networks, even home networks, are filled with tons of different network protocols that manage information flow. In this section, I'm going to briefly describe a few important ones you should be familiar with when building a home network.

TCP/IP: The backbone of the Internet

Transmission control protocol/Internet protocol, or TCP/IP, is the protocol that most computers use to talk to each other on home networks and over the Internet. Computers use TCP/IP to negotiate when to start data transfers, to determine which computer is sending and/or receiving data, to verify that data was received correctly, and to know when to end data transfers. TCP is responsible for the data delivery; IP is responsible for identifying the computers that are talking to each other. Networked computers have an IP address that either identifies them on the Internet (a public IP address) or identifies them on a private network, like a home network (private IP address).

Managing IP addresses is extremely important for home networks. There are multiple ways and considerations for managing private IP addresses, which we'll discuss in the planning and building your home network sections of this book.

Wi-Fi protocols: The standards that give our mobile devices life

Wi-Fi protocols are typically identified by their Institute of Electrical and Electronics Engineers (IEEE) designation. They all start with 802.11, which is the part of the IEEE local area network standard that specifies protocols for implementing wireless computer communication. Below is a table of current Wi-Fi protocols and some of their characteristics:

Protocol	Frequency	Maximum Data Rate
802.11ax (Wi-Fi 6)	2.4, 5, or 6 GHz	9.6 Gbps
802.11ac (Wi-Fi 5)	5 GHz	3.5 Gbps
802.11n (Wi-Fi 4)	2.4 or 5 GHz	600 Mbps
802.11g	2.4 GHz	54 Mbps
802.11a	5 GHz	54 Mbps
802.11b	2.4 GHz	2 Mbps

Devices can support more than one protocol. For instance, a device labeled 802.11a/g/n/ac supports 802.11a, 802.11g, 802.11n, and 802.11ac.

DNS: Putting a name to a number

Domain Name System, or DNS, is the system that assigns names to IP addresses. Imagine having to remember 172.217.164.110 instead of google.com, or 64.233.177.190 instead of youtube.com. DNS maps names to IP addresses to make services like websites much easier to remember. DNS also allows the IP address to change (e.g., moving the website to another server) without having to tell everyone the new IP address. Not having DNS would be like having to remember a person's social security number (or equivalent) instead of their name.

DHCP: The easy way to assign a lot of IP addresses

Devices can have static IP addresses that are configured manually on each device, or they can be assigned dynamic IP addresses automatically. Dynamic Host Configuration Protocol (DHCP) is a network management protocol used to automate configuring dynamic device IP addresses. As networks get larger, it's easier to manage IP addresses using a DHCP server. Usually, your router functions as the DHCP server on your network.

SMB/CIFS: Sharing files with other computers (and people)

One important reason to create a computer network is to share files with other computers and people. It's great to be able to access private photos and home videos on multiple computers in your home. SMB/CIFS (Server Message Block/Common Internet File System) is a very popular protocol for sharing files on a home network. You may hear of SMB or CIFS file sharing. CIFS is just a particular implementation of SMB; however, since it is from Microsoft, it is very popular and worth mentioning along with SMB.

HTTP/HTTPS: The protocol that powers the web

HTTP, which stands for Hypertext Transfer Protocol, is used by web servers to produce all of the wonderful web pages you view while browsing the web. It is the foundation of any data exchange on the web. Using web browsers to access the web is one of the most common home network activities and thus important for your home network planning. HTTPS is a secure version of HTTP, utilizing encryption to keep information transfer over the web secure.

VPN: Connecting securely to remote networks

Virtual private networks, or VPNs, allow users to connect securely to a remote network. For home networks, a common use is to allow access to your home network resources (e.g., printers, web servers, smart home controllers, and file servers) from any Internet connection around the world.

Common Devices That Connect to a Home Network

Why is a home network important? The main reason is so that the devices on it can easily connect to the Internet and/or each other. You may think the types of devices that connect to a home network are obvious. Most people think of their computer or laptop, mobile phone, and tablet. In today's world, people are also familiar with video streaming devices also connecting to their home network. Here are a few more devices you may not have considered:

- **Printers** – Most modern printers are also part of your network. This allows you to send a to-print command from other devices on your network straight to the printer without connecting directly to it.

- **Audio/Video (A/V) Receivers** – Modern A/V receivers connect to your home network so you can cast audio to them and stream audio from your favorite audio service.

- **Smart TVs** – It's getting harder and harder to buy a new TV that isn't also a smart TV—a TV that has apps for streaming services like Hulu, YouTube, and Disney+. These require a network connection to work properly.

- **Video Cameras** – Security cameras and video doorbells have risen in popularity over time. Many of these cameras utilize Wi-Fi or PoE to connect to your network.

- **Smart Devices** – Smart home technology continues to become ubiquitous. Smart devices like smart speakers (Amazon Echo, Google Nest Audio), smart switches, smart plugs, smart appliances, smart locks, smart bulbs, etc., all use a connection to your home network to utilize their smarts.

How Much Does It Cost to Build a Home Network?

Before building out your home network, it's important to understand the costs involved so that you can plan and budget accordingly. To get started on a home network is relatively inexpensive. Even a fully functional, stable, and secure home network doesn't have to break the bank. That said, it's also easy to spend a lot of money on a home network. The needs of different sizes of home networks can vary greatly, so we'll compare three different home network scenarios:

- A small home network with only a few devices
- A medium-sized home network with 20+ Wi-Fi devices
- A large home network for a house with multiple floors and 50+ devices

We'll also discuss a few special considerations that may add on to the cost of your home network. Note, this section assumes you are doing all the home network installation and configuration yourself, so no labor costs are considered in these estimates. Also, these are all only hardware costs. You can assume you'll be paying US$50–$130/month for your

Internet service. Later in this book, we'll discuss the specific hardware that will work for you.

Small Home Network Cost

Most home networks start small. They have fewer than 20 devices. Most devices are probably wireless, while a few are wired. In this scenario, you just need a modem for connecting to your ISP's Internet service, and an up-to-date Wi-Fi router. The cost for a small home network isn't too high:

- A modem: US$50–$200
- A Wi-Fi router with integrated gigabit ports: US$50–$100
- Total cost for a small home network: US$100–$300

The cost of modems and routers can vary a lot based on the type of Internet service you have and the features you want in your router. Your ISP will likely rent you a modem and maybe even a modem/router combo unit. In almost all cases, you'll be better off buying your own modem and router separately. Modem/router combos offered by ISPs are notoriously limited in feature set and more susceptible to hacking. Also, over time, renting the modem will cost you more than buying.

Medium-Sized Home Network Cost

As you start to add more and more devices to your network (like streaming sticks, smart home devices, and more mobile devices), you may start having problems with your Wi-Fi, and your family members may become frustrated with disconnections and slowdowns. In this case, you need to add some hardware to help handle the load of all the devices you now have on your network. Here are some additional costs you may incur:

- A 4-8 port gigabit network switch: US$20–$50
- A Wi-Fi access point: US$50–$100
- Additional Ethernet cables: US$20–$40
- Additional cost for expanding your home network: US$90–$190
- Total cost so far: US$190–$490

Large Home Network Cost

Your smart home has grown larger than you imagined it would. You now have well over 50 devices on your network, most of them on Wi-Fi. Maybe you've decided you want to turn your home into a smart home and anticipate having many more devices on your network, and you want to make sure everything is responsive. You've also noticed some weak Wi-Fi spots in your house. These are some additional costs you may incur:

- 1-2 more gigabit switches for additional wired devices: US$20–$200
- 2 additional Wi-Fi access points or a mesh Wi-Fi system: US$100–$400
- Additional Ethernet cables: US$30–$60
- Additional cost for expanding your home: US$150–$660
- Total cost so far: US$340–$1,150

Other Home Network Cost Considerations

The size of your home network isn't the only factor in determining your home network costs. Here are a few other things to consider:

- **The size of your home** – If you have a large home, or a multi-story home, you may want to consider a mesh Wi-Fi system or add wireless access points, even if you don't have a large home network. It's hard for a single wireless router to cover an entire house if it is large, especially if it can't be positioned centrally in the house. A mesh system or wireless access points will allow the Wi-Fi signal to be strong throughout your home.
- **Use of Power over Ethernet (PoE)** – If you want to power some of your devices, such as video cameras, using PoE, then you'll increase your costs. You'll

need to buy PoE-compatible switches, which cost more than regular switches.

- **Advanced switch functionality** – If you want advanced switch functionality, such as link aggregation and VLANs, you need to buy *managed* switches instead of regular switches, which also cost more. We'll discuss link aggregation and VLANs and why you might be interested in these technologies later in this book.

- **Retrofitting your home with wired connections** – Let's say you really want the speed and reliability of wired connections at various locations in your home but don't have Ethernet connections in those locations. You will need to add the cost of running Ethernet cables to new locations from your router. How much this costs depends on the specifics of your house and how much cable you want to run. I can tell you from experience, however, that it is usually worth it. It's also something to consider if you ever get the chance to design and build a new home.

- **Backup Internet access** – Some people, like me, are obsessive about having reliable Internet connections. You really don't have much control over your ISP's reliability, but you can have a backup Internet connection to fall back on. This will cost you more in monthly service fees, and you'll want to buy a router that supports multiple Internet connections. You guessed it—these routers cost more.

What I Do: Cellular Backup Internet

I wrote earlier that I'm obsessed with having reliable Internet service. Although people complain a lot about their Internet connections, the truth is most ISPs (Internet Service Providers) do a good job of staying up. But for the paranoid, or for those who have shaky ISPs, cellular backup Internet service can save a lot of headaches.

I live in an older neighborhood with lots of trees that like to fall when the weather gets bad. So every year, we lose power. Our power and Internet service lines are mostly above ground. One winter, we were without power for many days, and without Internet service for even longer. So I decided I'd look for an inexpensive Internet backup.

I had backup Internet for a month when I switched over to CenturyLink, because I kept Xfinity going just to make sure CenturyLink was going to be stable enough. We had one outage for a couple of hours (in the middle of the night), and Xfinity kicked right in.

Why not just keep Xfinity as my backup? When power or Internet goes out for long periods, it's usually because of damage to a utility pole. CenturyLink and Xfinity use the same utility poles around here, so chances are they'd both be out at the same time. I wanted something that would be more fault tolerant. That led me to cellular backup.

I purchased a Netgear 4G LTE modem, which works with all the major carriers. It uses a standard SIM card to get cellular Internet and also has an Ethernet port for connecting to the WAN (Wide Area Network, or Internet) port of the router. Then, I used a pay-as-you-go cellular plan.

My router is set up to switch to the cellular connection whenever my regular Internet goes down. It's not the fastest, and I have to pay a lot if I use significant data, but it's good to have this option in a pinch.

How Do I Do That? Retrofitting Your Home with Ethernet

If you decide that you need wired Ethernet connections in places in your home that don't have them, you'll have to do some Ethernet retrofitting. This can be a big project; however, in many cases, it doesn't have to be. If you have a crawl space, or an attic, you can often easily run wire yourself to different rooms without doing major drilling or fishing of wires. You can drop wires down a wall from the attic or push them up a wall from a crawl space.

You'll want to buy wall plates and learn to use an Ethernet crimping tool to connect the wires to the wall plates. You'll also want a paddle bit or a hole saw to make the holes to install the wall plates. Fortunately, these aren't expensive tools and supplies. They are also easy to learn to use. Buy the best Ethernet wire you can afford, and consider running multiple lines to whatever room you have in mind. If you have Ethernet connections for each device, you won't have to buy a switch for that room later.

If all this seems like too much, an electrician or a handyman should be able to easily get the job done. The point is that retrofitting your home with Ethernet cables is usually a straightforward project.

Key Takeaways

❑ A home network connects all the computers in your home so that they can communicate with each other and share an Internet connection.

❑ A home network is important because of all the things it allows you to do, including the following:

❑ Share an Internet connection with devices in your home

❑ Send a command to a printer in your home from multiple computers

❑ Stream video to your Fire TV, Roku, or similar streaming device

❑ Set up a Wi-Fi network so you can use Wi-Fi-dependent devices (video cameras, smart speakers, tablets, etc.)

❑ Share files across computers

❑ Back up your important data to the cloud

- ❑ Work from home with a stable Ethernet connection

- ❑ Use most smart home devices (smart switches, smart plugs, smart lights, etc.)

❑ Important home network devices include the following:

  - ❑ Modems

  - ❑ Routers

  - ❑ Firewalls

  - ❑ Switches

  - ❑ Wireless access points (WAPs)

  - ❑ Mesh Wi-Fi systems

  - ❑ Wireless extenders

  - ❑ Network cards

❑ Different Internet connection types include the following:

  - ❑ Dialup

  - ❑ ADSL

  - ❑ Cable

  - ❑ Fiber

  - ❑ Cellular

  - ❑ Satellite

❑ Important home network protocols include the following:

  - ❑ TCP/IP

  - ❑ DNS

  - ❑ DHCP

❏ SMB/CIFS

❏ HTTP/HTTPS

❏ Home networks can be relatively inexpensive, but costs can add up quickly.

PART 2

BUILDING YOUR HOME NETWORK FROM SCRATCH

Evaluate Your Home Network

If you are reading this book, chances are you already have some type of home network. You probably have a modem and router, a computer or two, and some mobile devices. Many people will have more than that. As the first step in building your ideal home network, you should take the time to evaluate the state of your home network. Your current home network will form the foundation of your new home network and inform your needs. In this section, we'll step through a few areas you should evaluate. I will ask you a few questions as we go, and you should write down the answers. They will help you in the remaining steps of building your home network.

Is Your Current Internet Service Fast Enough and Stable Enough?

The bandwidth of your Internet service determines the quality of your connection to all Internet services outside of your home. Streaming video, web surfing, cloud file collaboration, video conferencing, and so much more depend on your Internet connection. Do you or your family become frustrated with frequent Internet outages? If so, you may want to consider other ISP options in your area.

Evaluating the bandwidth (speed) of your Internet connection takes a bit more work. How much speed you need depends on how you use the Internet, and how many people and devices use the Internet in your home. Primarily, you need to understand your Internet download needs. Download bandwidth needs are determined by the resources you pull into your network from the Internet (streaming services, gaming, torrenting, etc.).

First, go to Speedtest.net[1] from a wired computer on your home network and test your Internet speed at different times of the day. Make sure no other devices are doing any significant downloading or uploading to the Internet (streaming videos, downloading large files, etc.) at the same time. Are they the speeds you were expecting based on the service you purchased from your ISP? If not, make sure to complain to your ISP.

Second, you should determine your bandwidth needs. For example, Netflix recommends 5 Mbps for each high-quality HD Stream and 25 Mbps for each Ultra HD (4K) stream. Below are some common bandwidth needs for home network activities:

- HD video streaming: 5–25 Mbps per stream (downloading)
- Music streaming: < 1 Mbps per song (downloading)

---

[1] https://www.speedtest.net/.

- Online gaming: 3–6 Mbps download / 1–3 Mbps upload per person
- Web/social media: 3 Mbps download / 1.5 Mbps upload per person
- HD Video Call (e.g., Zoom): 1–4 Mbps up-load/download per person

The amount of bandwidth your home needs depends on the number of people and their regular usage of the Internet at home. For example, let's say it's common in your home for a couple of people to be streaming a video (~15 Mbps), for two more to be browsing the web/social media (~6 Mbps), and for someone else to be gaming online (~6 Mbps) and streaming music in the background (~1 Mbps)—all at the same time. Adding up 15, 6, 6, and 1 gives you up to 28 Mbps, which is the minimum download bandwidth needed to support these activities. You should always add at least a 30 percent overhead to your calculated minimum. The speeds ISPs quote are maximum figures that aren't always available. Often, there are other devices on the network that are steadily using small amounts of Internet bandwidth (e.g., checking for and downloading updates). Keep in mind that if you frequently view and/or send videos, pictures, and other files via social media, your needs may be higher than the average 3 Mbps figure above. Also, there are often spikes in the rates needed by these services. For example, an online game may average 3 Mbps but have certain spikes where it needs 10 Mbps to function properly. In the example I gave earlier, I'd recommend you to actually get around 40–50 Mbps to cover your needs.

If you constantly download content (via torrents and/or Usenet, for example), then you already know that you have high bandwidth needs. Ultimately, it comes down to how fast you want those downloads to happen and how often you download.

You may wonder about upload bandwidth. Examples of activities that require more upload speed include VPN connections, video calls, and some online games. Uploading usually requires significantly less bandwidth than downloading does. Still, if you are going to be doing a lot of activities that require upload bandwidth, you should calculate your upload bandwidth needs. Many ISPs (like cable) offer an upload bandwidth that is significantly lower than the download bandwidth. Make sure to check the speeds of both when shopping for an Internet connection service package.

If your current Internet bandwidth exceeds your Internet needs and is reliable, then you are probably fine and don't need to look for a new package or ISP. If not, it is time to shop around. We'll discuss finding the right Internet provider after we're done evaluating your home network.

Is Your Current Networking Hardware Good Enough?

Let's start with your modem, which connects your home network to your ISP's network, and thus to the Internet. Are you renting your modem from your ISP? When was the last time you upgraded your modem? If you are renting your modem from your ISP, you should strongly consider buying your own. Generally, you'll break even on the cost of your own modem in less than a year. If you haven't upgraded your modem in many years and you are not getting the Internet speeds you think you should be getting, you should investigate whether your modem supports the speeds you are paying for.

Next, let's move on to evaluating your router. Your router has an enormous impact on the security, speed, and capabilities of your network. How old is your current router? Do you have to reboot it often? When was the last time it received a firmware update? The answers to these questions will help determine whether you need to buy a new router in order to improve your home network.

Lastly, let's evaluate your switches if you have any. Switches usually just work—without any fuss or maintenance. The main thing is to make sure they are at least gigabit switches. We are transitioning to more and more devices that can exceed older and slower "fast Ethernet" (100 Mbps) switches. Fast Ethernet, which was fast for its time, doesn't seem to be appropriately named now. If your switches are slower than gigabit, they may become a bottleneck in your network.

Does Your Wi-Fi Frustrate You?

Wi-Fi devices have proliferated over the last few years as has our dependence on them. Few things frustrate family members more than unstable or slow Wi-Fi connections. Make note of places in your home where Wi-Fi doesn't work as well as you want. Do you have constant dropped connections or slow Wi-Fi? Does your Wi-Fi network support the faster, more stable protocols like Wi-Fi 5 (802.11ac) and Wi-Fi 6 (802.11ax)? Do you have a guest Wi-Fi network?

What Is the Environment Like in and around Your Home?

Some Wi-Fi problems may be related to the environment around your home. For example, if you live in an apartment complex, you may have a lot of competing Wi-Fi signals that can cause issues. Some apartment complexes supply the Wi-Fi for their tenants, meaning you don't have to worry about your Wi-Fi signal. Even if you live in a house, you may have Wi-Fi interference from close neighbors, which you have to deal with.

The location of your home isn't the only environmental factor. The materials your home (or your apartment building) is made of can affect how far wireless signals can travel.

Concrete, brick, and thick timber walls can significantly impair Wi-Fi signals. Other electronics inside of your home can also affect your Wi-Fi signals. Microwaves, baby monitors, and even other Bluetooth devices are capable of disturbing your Wi-Fi.

Whether you own your home can play a role in your home network design as well. For example, you may have very limited options, or no options at all, for running wired Ethernet in your walls to additional rooms. These are environmental conditions to consider when designing your home network.

What Devices Are Currently Connected to Your Home Network?

What devices are connected to your network, how they are connected, and where they are located are important pieces of information to document as you plan your home network, and to keep track of as your network grows and changes over time. How many devices you have, especially Wi-Fi devices, informs the type and number of devices you need to handle your network load efficiently. It also tells you what types of Wi-Fi protocols you need to support. Additionally, knowing all of the devices on your network is paramount in detecting if and when rogue devices connect to your network. If you know all the devices that are supposed to be on your network, you'll know any device that you don't recognize is not supposed to be on your network. Detecting rogue devices on your network is discussed in part 3 of this book.

What Isn't Working Well on Your Network?

The final aspect of your home network that you should evaluate is how well it is working. Do some of your Wi-Fi devices have problems staying connected? Is your Internet speed too slow or not able to handle all the devices it needs to? Do you reboot your router from time to time because it freezes? You should note all the frustrations your home network users have experienced. Addressing these issues should be part of your plan to rebuild your home network.

How Do I Do That? Identifying All of Your Home Network Devices

Earlier in this book, I mentioned inventorying all of the devices connected to your home network. This may be a difficult task, even if you have great memory. There could be a lot of devices on your network to remember, and a roommate, a spouse, or a child may have added a device without your even knowing. Fortunately, there are a few good ways to find out all of the devices connected to your home:

- **Check your router.** This is probably the easiest method. Most routers allow you to look up all the devices on your network. Where this information exists on your router varies. Generally, look for a status or DHCP page.

- **Scan your network.** There are many different tools that can scan your network. First, you need to find out the subnet (IP address range) of your network. For Windows, you can look into tools like Advanced IP Scanner, SolarWinds IP Address Manager, and NMAP. You can use NMAP on Apple and Linux devices as well.

What I Do: Evaluating My Home Network

Once a year, I take stock of my entire home network. I keep spreadsheets of all my wired and wireless devices, including their IP address, their hostname, what Wi-Fi protocols they support, and where they are located. I also have a network diagram of all my rooms that includes all the devices in each room. I try to keep this updated throughout the year; but once a year, I make sure it is updated. It helps me see whether certain parts of my home are overloaded with Wi-Fi devices and whether I need to make any upgrades to the equipment.

Create Your Home Network Goals

Now you've documented a baseline for how well your home network serves your needs and what network devices you are currently supporting. It's time to define the capabilities you want your home network to have. This information will inform what equipment (if any) you need to buy, how you configure your home network, and how you protect your home network from all the bad actors out there. Let's get started! You just need to be able to answer the following questions:

What Do You Want to Improve?

If you followed the home network evaluation section of this book, you probably already have this answered. Most people want to address pain points like spotty Wi-Fi and Internet speed. Others, however, want to improve their home network security. Make a list of all the things you want to improve about your home network.

How Many Devices Do You Want Your Home Network to Support?

You already know how many devices you have on your network now (if you don't, check out the earlier section about how to find all of the devices on your network). Next, think about what types and how many devices you plan to add over the next 3-5 years. Are you going to add security cameras, smart speakers, streaming devices, or other similar devices? Do you plan to get a new tablet, a laptop, or a PC for a family member? How about a smart thermostat, a smart garage door opener, or a smart video doorbell? All of these devices will use your network. Having a general sense of the number of wired and wireless devices you'll need to support will determine what hardware you'll require.

Do You Need Network Connections Outside of Your House?

If you live in a house, you probably also have a yard. Do you want to be able to use mobile devices in your yard? Do you have a separate garage or workshop that also needs network access? This may lead to additional wiring or access points that you need to buy to extend the range of your network.

Do You Want to Connect to Your Home Network Remotely?

Something most people building home networks don't think about is being able to connect to their home networks remotely. Connecting remotely to your home network can be done in many different ways. Maybe you just want access to some files you keep on your network. You may want to control your smart home devices or take a peek at your security cameras. You may want full access to all of your home network computing resources like PCs, printers, web servers, etc. Some of these scenarios can be solved with cloud connections to your devices and storing information in the cloud. Other scenarios would be best solved by having a remote desktop connection set up. Still, others may require you to set up a full VPN connection to your home network. Think about your needs, and that will determine the solution.

Select Your ISP and Home Network Hardware

I know this may seem like a lot of pre-work and planning to finally get to the point where you are buying the hardware for your home network. Trust me when I tell you it is better to do some research and define your goals before jumping into something as important and potentially complicated as building or improving a home network. Just remember, failing to plan is planning to fail. Here's how to choose your ISP and all the hardware you'll need.

How to Select Your ISP

Your home's Internet connection is your network's gateway to all the wonderful services on the web and in the cloud. Thus, your selection of an ISP is critical to the quality of your home network experience. You have already determined whether your current ISP is good enough in the home network evaluation section of this book. If it isn't good enough, or you just want to improve or re-evaluate your options, here is some advice on how to select your ISP:

1. Make a list of the ISP providers and packages available to you that fit your upload and download bandwidth requirements. You can usually find a list of ISPs in your area by going to HighSpeedInternet.com[2].

2. Compare the speeds, pricing (including installation fees, equipment prices, and early termination fees if you are leaving your current ISP), and read reviews of their service reliability in your area. Make sure to consider data caps if you think you'll be streaming a lot of data (lots of video streaming, video conferencing, gaming, torrenting). Some providers offer plans without data caps. If you need a plan without a data cap and you can't find one, start looking for business class service. They usually don't have data caps.

3. Select the plan that best meets your speed, data, and budget requirements. Remember to check reviews to make sure the service is reliable.

How to Choose Your Router

Your router has an enormous impact on the security, speed, and capabilities of your network. Choosing your router is

---

[2] https://highspeedinternet.com.

one of the most important home network decisions you'll make. If your current router is reliable, up to date, and meets your needs, you are already done. If it doesn't, you should take the following steps to choose a new router:

Determine how much space you need to cover

Most people use their router as a wireless access point. If you do, you'll want a wireless router that provides your entire home with stable, fast Wi-Fi. Consider at least two factors when determining the range:

1. What protocol(s) do you plan to use? Some, like 2.4 GHz protocols (e.g., 802.11n, 802.11g), cover more range and are supported by more devices. Others, like 5 GHz protocols (e.g., 802.11ac, 802.11n can be 5 GHz as well), cover less range but are faster and less prone to interference. The latest protocol is Wi-Fi 6 (802.11ax), but it will be a while before most devices adopt it.

2. Where in your house will you place your wireless router? Generally, the best place for whole house coverage is in the center of your house, but few peo-ple place a router there. If it is at the edge of a floor or house, it has less chance of covering the house.

Wi-Fi range for routers varies greatly. A good Wi-Fi router will cover 2,000 square feet easily. Some can cover up to

5,000 square feet. These are three key factors that can impact Wi-Fi range:

1. **Signal interference** – If you live in an area flooded with Wi-Fi signals (e.g., an apartment complex), you'll have interference that restricts your Wi-Fi signal stability and range. Microwaves, cordless phones, Bluetooth speakers, and even baby monitors can also cause signal interference.

2. **Walls and surfaces** – Materials like concrete, metal, and plaster are the worst for blocking Wi-Fi signals.

3. **Router placement** – I mentioned earlier that a router placement affects the signal range. Most routers are omnidirectional; in other words, they send and receive signals in all directions. Placing them next to concrete or brick walls will limit their range. Also, the higher up on the wall the better the range of the signal.

If you have a large home, one or more factors limiting your Wi-Fi signal range, or lots of devices, you should consider buying a router and supplementing it with additional wireless access points. Companies like TP-Link and Ubiquiti sell wireless access points specifically made to expand wireless coverage. Another option, especially if you don't have a wired network connection to connect wireless access points back to the router, is investing in a mesh router system (e.g., Google Nest Wi-Fi and Eero).

Determine your budget

I recommend that you budget at least $100 for a good Wi-Fi router. You can get away with less. However, if you can get to the $100 mark, then there is a significantly lower chance you'll be disappointed with your purchase. I'd recommend saving up until you can reach that point, because the wireless router is an important piece of equipment for your network. If you are willing to go up to $200, you should be able to get all the features you are looking for.

Pick a wireless router

Seems simple, right? Here are a few things to look for and some to avoid:

Features to look for:

❏ Support for Wireless AC (Wi-Fi 5). There's no point in buying a router that doesn't support this protocol. It's faster and more stable than the protocols that preceded it, and most mobile devices made in the last few years support it. Wi-Fi 6 would be a bonus and will help you in the future.

❏ Supports MIMO. MIMO (multiple-input and multiple-output) allows the radio in your router to connect to multiple devices without slowing down each connection when compared with wireless routers without MIMO. SU-MIMO (single user) is good, and MU-MIMO (multi-user) is even better.

❏ Quality of Service (QoS). This allows your router to specify the types of Internet traffic (e.g., video streaming, gaming, etc.) that have priority over other types of traffic. QoS can also be used to give priority to certain devices (e.g., your mobile phone or PC) over other devices.

❏ Supports at least WPA2 AES security for Wi-Fi. Any older standard isn't secure.

❏ Gigabit Ethernet ports. The wired connections to your router should also be fast.

❏ Guest Network. A guest network is an essential part of Wi-Fi security. I'll explain this in more detail in the security section of this book.

❏ Consistent firmware updates. You should be able to set up automatic security updates or at least be able to receive notifications or check for updates from the router interface.

A Couple of Things to Avoid:

❏ ISP's provided model. Chances are this model is bundled with a modem, making it more of a single point of failure from a security and reliability standpoint. Renting this unit over time will probably cost more than buying one, and these routers are often fraught with security issues.

❏ Older routers. Get one that is new or at least receives frequent and recent security updates.

Read reviews before purchase

You may have narrowed it down to two or three routers. Use professional reviews (CNET; Tom's Guide; or my site, HomeTechHacker) and user reviews (Amazon, Newegg) to get opinions and learn more about the router to help you make your decision. This is the way to get real-world answers if you have questions about the router.

What I Do: Using pfSense as My Router

A few years back, I became frustrated with off-the-shelf routers. There seemed to be many security vulnerabilities and not enough firmware updates. I decided to build my own router using pfSense software. pfSense is free for home use and has tons of features that are found in enterprise-class routers. It has all the features I need, is very secure, and has served me well for many years.

These days, there are better choices for routers, and pfSense isn't for everyone. However, if you are feeling a little adventurous, give it a shot.

How to Choose Your Switches and Cabling

With where technology is going, you should try to achieve at least gigabit speeds in your home. The more bandwidth you have, the more future proof it is and the less chance there is for bottlenecks. You can achieve this by making sure all wired networking equipment (switches, routers, access points) and all wired clients (desktop computers, laptops, streaming devices) support gigabit networking. It's not the end of the world if some devices support only fast Ethernet (100 Mbps). That's plenty fast for most uses. But you want to be prepared for the future.

Switches

When looking for a switch, first you need to figure out how many you need and how many ports each switch should have. Count the rooms you want to have multiple Ethernet devices in and decide on how many Ethernet devices you plan to have in each room. This will determine how many switches you need and how many ports each switch should have. Remember that your router will likely have a switch built into it. Also, try to plan for the additional devices. It's better to have one switch for all your connections in a room. Sure, you could just daisy chain another switch in the future, but doing too much daisy chaining of switches can lead to network performance issues.

Next, you need to consider what features your switches need. First, make sure the switch supports at least gigabit speeds. Second, determine whether you need a managed or an unmanaged switch. You may be interested in a managed switch if you want fine control of traffic prioritization, the ability to link multiple ports together to increase the bandwidth of a network connection (link aggregation), the ability to create VLANs (for segmenting your network, more on that in part 3 of this book), or other more advanced features. Otherwise, an unmanaged switch will work fine and save you money. Most people setting up home networks do fine with unmanaged switches.

One final consideration for your switches is whether you want to use power over Ethernet to power some of your devices. If so, then you need to purchase PoE-capable switches, where you want to connect PoE devices.

Cables

You can support gigabit speeds and make your home more compatible with future speed technologies by using at least Cat 5e cabling and preferably Cat 6. Cat 5e/6–rated cable will support gigabit speeds for longer distances and more reliably than older standards like Cat 5. While recabling a home can be hard, if you have the opportunity to choose your cable, go with at least Cat 6 or even Cat 7.

One other consideration for the wired part of your home network is that you should use as many wired devices as possible. Although Wi-Fi standards keep improving, the fact is that wired connections are usually more reliable. Try to save your wireless bandwidth for devices that truly need it (e.g., mobile devices and devices that don't have wired networking).

If there is an area in your home where you can't easily run an Ethernet connection and the Wi-Fi connection is spotty, you can consider powerline or MoCA adapters. As mentioned earlier in the book, powerline adapters allow you to use your home's electrical wiring as a network connection, and MoCA adapters take advantage of your home's coaxial wires.

How to Choose Your Access Point(s)

Recall, that for our purposes, a Wi-Fi access point is a device that connects back to your router via a wired connection (or to a switch that connects to your router) and provides a Wi-Fi signal for your home network. The first consideration when choosing an access point is whether you even need one! How well does the Wi-Fi coverage in your house work? If you have a smaller house, or a centrally located Wi-Fi router, you may not need an access point. You also don't need one if you have a mesh Wi-Fi system. Your mesh Wi-Fi system should provide all the coverage your home needs. You'll be better off buying another satellite for your mesh system than an access point if you need more coverage.

If you've decided that you need better Wi-Fi coverage, then you need to make sure you have wired connections in places where you'll consider placing access points. If you don't have wired connections to plug them into and you don't, or can't, run more wire, then access points won't work for you.

Once you've determined you can benefit from an access point and that you have the infrastructure to install it, it's time to choose one. You should look for many of the same Wi-Fi features that are important for your Wi-Fi router: supporting at least Wi-Fi 5 (802.11ac), and preferably Wi-Fi 6 (802.11ax); MU-MIMO; having guest network capabilities; and receiving regular firmware updates. Other things to look for include the following:

❏ Number of antennas (spatial streams) – At a minimum, a Wi-Fi device (like your cell phone) will have at least a 1x1 radio, meaning it has 1 radio chain and it can support a single stream in and a single stream out. As I mentioned earlier, you want a MIMO (remember, that stands for multiple in, multiple out) device, which means you need to have more than one radio chain. Look for one that is at least 2x2. Even better are 3x3 or 4x4 chains, especially if you have a lot of Wi-Fi devices.

❏ Dual band support – Chances are, you have devices on your home network that are 2.4 GHz. Even though it is imperative that you buy a router that supports the newer 5 GHz protocols, you still want it to support all of your Wi-Fi devices. So, it needs to support both 2.4 GHz and 5 GHz. It may be specified as 802.11 a/b/g/n/ac or 802.11 a/b/g/n/ac/ax. The good news is that if it supports Wi-Fi 5 or Wi-Fi 6, it is most likely a dual band access point.

❑ Central management software – This is software that allows you to configure and manage multiple access points from one interface. If you get an access point that is the same brand as your router, it likely can control that too. It is nice to not have to configure all the access points separately. Central management software greatly improves the interoperability of the access points while reducing the effort required to configure and manage them.

Other Network Equipment to Consider

We've gone through the most common network equipment you'll need to have. Here are some other network devices and the reasons you may want them:

❑ Wi-Fi extender – If you have a small area in your home that doesn't have a good Wi-Fi signal, and you don't have and can't run wired Ethernet to that area, then you can't use a Wi-Fi access point. A Wi-Fi extender is a good alternative.

❑ Security appliances – These are devices that connect to your network with the sole purpose of protecting you from threats like malware and hackers. These devices go beyond the security of most routers. While these devices can add peace of mind to your home network security, they can be difficult for the average home user to configure. Examples of these devices targeted for home users include the Fire-walla, the Bitdefender Box, and the Zyxel Next Generation VPN Firewall.

❏ VPN appliances – These are appliances that connect to your home network and allow for VPN connections to parts or all of your home network. They often have many of the features of the security appliances mentioned earlier and can also be difficult to configure. Many routers come with built-in VPN capabilities. Examples of these devices include the Zyxel Next Generation VPN Firewall and the Sophos XG 86 Next-Gen VPN Firewall Appliance.

❏ Travel routers – These are usually pocket-sized Wi-Fi routers that you can take with you when you travel. Why would you do that? They can simplify the process of connecting and managing devices on public Wi-Fi networks like at hotels and airports. You keep your mobile device configured to connect to the travel router, and then you configure the travel router to connect to the public network. This can give you additional privacy and make it easier if you have multiple devices that need to connect to a new network. Also, some hotels still offer for free only wired Ethernet connections. This device can turn that wired Ethernet connection into a wireless one.

❑ Wi-Fi bridges – A Wi-Fi bridge turns a Wi-Fi connection into an Ethernet connection. It is kind of the opposite of a Wi-Fi access point. You'd want to use one of these when you have a device that accepts only a wired connection in a place where you don't have Ethernet. Another use is for long-range connections. Let's say you have a barn or a shop that is far away from your house, and you don't want to run Ethernet underground to it. You can buy a long-range Wi-Fi transmitter and bridge (sometimes called a point-to-point wireless bridge) to add a wired connection to your barn or shop, and then set up the network (with switches, access points, etc.) just like you would in your home.

❑ Powerline Ethernet – As I mentioned previously, powerline adapters allow you to run wired Ethernet through your home's electrical wiring. This may be a good alternative to a Wi-Fi bridge or Wi-Fi range extender.

❑ MoCA adapters – You'll need these if you want to try and use your home's coaxial wires to transport signals across physical wires in your network. MoCA adapters may be another good alternative to a Wi-Fi bridge or a Wi-Fi range extender.

As you are selecting and purchasing your home network gear, remember your evaluation of your current home network. You may not need to buy as much as you think.

How Do I Do That? Making Internet Usage Safer for My Children

One of the most common questions I'm asked is how to protect children from inappropriate content on the Internet. The best and most important step is to educate your children how to distinguish between what is appropriate and what is inappropriate and how to recognize bad content and behaviors on the Internet. I know that answer isn't enough for many families.

Other options include blocking inappropriate sites and ads on your children's devices. Many routers come with these features as do the security appliances I mentioned earlier. Another option is to use a service like OpenDNS to block specific websites. OpenDNS works by blocking sites at the DNS level. You specify which domains (or categories of domains) are allowed and not allowed. The technical part of OpenDNS is that you have to know how to assign DNS servers to your computers. This isn't too hard to learn. OpenDNS has a free level of service that should work for most people.

Additionally, if you have Android phones, you can use a free service like Google Family Link. This service allows parents to control the apps, websites, and content levels of sites like YouTube for their children. It also allows you to specify the amount of screen time for Android and ChromeOS devices the children get each day, including how much time they can spend using each app. There are other similar paid services that work on both iOS and Android. The benefit of Family Link is that it can also work on web content on non-mobile devices.

Put Together and Configure Your Home Network

You've done all the research. You've completed all your planning. All of the network hardware you're going to use has been bought and is at your home. Now it is time to put it all together to have the home network you've always wanted. First, we'll start with your network infrastructure.

Make Your Planned Infrastructure Improvements

In your planning, did you decide to run Ethernet to additional places in your home? Do you plan to mount access points on walls? Do you want to run PoE network connections in your attic? Are you upgrading your wiring to Cat 6A? Replacing your PC network cards with gigabit cards? All of these improvements should be done before you start installing and configuring network equipment.

Test Your ISP's Speeds

If you've switched your ISP as part of this planning, or you simply haven't tested your ISP, do that first. You want to make sure you know what speeds you are getting. In this case, since you are rebuilding your home network, I recommend you plug a computer directly into the modem to test your speed. Make sure you've got your firewall running. You'd be surprised how fast hackers will start trying to access your computer. This computer should be one with a gigabit Ethernet card (or better) that you know has reliable network performance. You can use Speedtest.net. This will also serve as a baseline as you add more network components. You'll know what Internet speeds your network is capable of. There will be a couple of points throughout your network configuration where you'll want to test speed again to make sure you haven't introduced any significant slowdowns.

Configure Your Router

Aside from the planning, this is the most important step in building and configuring your home network. Let's work through configuring a router step by step:

Step 1: Decide where to place your router

For most people, it will need to be placed near the modem their ISP offers them. This, unfortunately, is usually on a side or even in a corner of the house. If it is a Wi-Fi router, and you are expecting it to supply a Wi-Fi signal throughout your home, placing it somewhere in the middle of your home would be ideal if possible.

Aside from a physical location, you want to place your router away from other radio frequency signals like microwaves, wireless phones, and baby monitors to minimize interference with the Wi-Fi signal. You want to place it as high up as you can to get the best signal. Lastly, you want to make sure to place it close to a power source and any computer or device (such as a switch or a computer) that has a wired connection to it.

If you are using a mesh router, make sure to look at the section about additional considerations for configuring it.

Step 2: Connect the router's Internet connection

Connect the Ethernet cable coming from your ISP-provided modem into the WAN port of your router. Next, connect a computer or a laptop to one of your router's LAN ports. Test that you have an Internet connection by going to speedtest.net again. You should still be getting close to the speeds you had with a direct connection to the modem. If you are not, try a different laptop or a PC. If that doesn't work, your router may not be fast enough to handle your Internet speed.

If you are using a modem/router combination from your ISP and don't want to use your own router, you can skip this step. However, if your ISP requires you to use a modem/router combination from them, and you want to use your own router, you'll need to find out how to put their modem/router in "bridge" mode. This mode will stop IP address conflicts between your router and theirs and allow your router to act as the DHCP server on your network. After putting their modem/router in bridge mode, you can connect it via Ethernet to your router's WAN port.

Step 3: Log in to the router and change default username and password

You never want to leave the default username and password on any Internet-connected device. It leaves an easy door to compromise your network security. I highly recommend that you change the default username and password before doing any configuration of your router.

Log in to the router, using a web browser on a computer directly connected to the router via a wired connection. You'll need to know the IP address of the router to log in. Unfortunately, the default IP address of various routers varies as do the default username and password. Sometimes you can find them on the bottom of the router. Otherwise, you can probably find them in the user manual. Below is a table of popular router brands and their default IP addresses, usernames, and passwords to help you. Some brands aren't consistent across all their router models. Usually, trying a combination of the popular IP addresses and usernames and passwords will get you in. Some routers provide an app for discovering and configuring the router. Once you log in, go ahead and change the default username and password.

Manufac-turer	IP Address	Default Username	Default Password
3Com	http://192.168.1.1	admin	Admin
Apple	http://10.0.1.1	root	alpine
Arris	http://192.168.0.1	admin	password
Asus	http://192.168.1.1	admin	admin
Belkin	http://192.168.1.1	admin	admin
BenQ	http://192.168.1.1	admin	Admin
DELL	http://192.168.1.1	admin	password
D-Link	http://192.168.0.1	admin	Admin
Linksys	http://192.168.1.1	admin	Admin
Netgear	http://192.168.0.1	admin	password
Netstar	http://192.168.0.1	admin	password
pfSense	http://192.168.1.1	admin	password
Samsung	http://192.168.0.1	admin	password
Sigma	http://192.168.0.1	admin	admin
Sitecom	http://192.168.0.1	sitecom	password
Sun	http://192.168.0.1	admin	admin

Synology	http://192.168.1.1	admin	Admin
Telco	http://192.168.0.1	telco	telco
Trendnet	http://192.168.1.1	admin	admin
TP-Link	http://192.168.1.1	admin	admin
Zyxel	http://192.168.1.1	admin	1234

Step 4: Check for firmware updates

Keeping your router firmware up to date is a key step in maintaining your home network security. After logging in, make sure you are running the latest firmware before continuing configuration. Usually, there is a screen in the router configuration that tells you the firmware version the router is running, when it was last updated, and whether there is new firmware to download.

Step 5: Configure DHCP

Configuring DHCP involves two primary steps: (1) picking the range of IP addresses you want to use on your network and (2) setting up a smaller part of the range you picked for your router to dynamically assign to clients. Your range of IP address choices is as follows:

Start Address	End Address	# of Individual IP Addresses in Range
192.168.0.0	192.168.255.255	65,536
172.16.0.0	172.31.255.255	1,048,576

10.0.0.0	10.255.255.255	16,277,216

These are the private IP ranges available for use. Keep in mind, you typically would pick not one of these entire ranges but rather a subset of the range. How do you pick which range to use?

I generally advise against using anything in the 172.16.0.0–172.31.255.255 range, because it is too easy for people to mistakenly overlap with one of the public 172.x.x.x ranges of IP addresses, causing problems. To keep things simple, I recommend that you choose 192.168.x.0–192.168.x.255 if you are going to have a smaller network (this range contains 254 *usable* IP addresses). As you saw in the router table earlier, many routers assume an IP address in the 192.168.1.x or 192.168.0.x range. These will work fine. If you want to use something more obscure, you can use 192.168.20.0–192.168.20.255 (also known as 192.168.20.0/24 or 192.168.20.0 with a subnet mask of 255.255.255.0). Just remember to make sure to assign your router an internal IP address in the range you choose.

If you find yourself needing more than 256 devices, then the simplest thing to do is to choose a range with a larger subnet mask. For example, if you were originally using or thinking of using 192.168.0.0–192.168.0.255 (192.168.0.0/24), and you need 500 IP addresses instead, you could use 192.168.0.0/23 (192.168.0.0 with a subnet mask of 255.255.254.0). This corresponds to the 192.168.0.0–

192.168.1.255 network, which contains a whopping 510 usable addresses.

Next, you need to decide how big of an address range you want to let your router hand out via DHCP (dynamic IP addresses) and how many addresses will be static (manually configured). Devices that have dynamic addresses can change IP addresses each time they connect to the network. Devices that have static IP addresses will have the same IP address every time they connect to the network. Any device that other devices on your network will regularly communicate with (like a file server or a printer) will be much easier to deal with if they have a static IP. To make an address range decision, you need to think about how many devices will be static on your network. If you have an IP address range of 192.168.0.0–192.168.0.255 and you plan to have 200 static IPs on your network, you may want to make your dynamic address range (the range the router hands out dynamically) 192.168.0.201–192.168.0.254 (192.168.0.255 is a broadcast address and not available for assignment in this subnet example).

I generally recommend assigning static IP addresses to as many devices as possible. It's easier to manage devices with IP addresses that don't change. Troubleshooting network problems is easier when you know the IP address of a device won't change. There are two ways to give devices static IP addresses. One is to use the devices' IP configuration to assign it a static IP. This configuration varies by device, and some devices don't have an interface for assigning a static IP. Another way to assign static IPs is to have your router reserve IPs for devices based on their media access control (MAC) address. A MAC address is a unique identifier assigned to each network interface. Most routers can assign and map an IP address to a MAC address. The tricky part is figuring out the MAC address. Sometimes, it is on the device or is printed on the box. Other times, you will need to connect the device to the network and look up its MAC address in the router or device configuration pages. This is the method I primarily use to assign static IP addresses. Technically, these are called static leases.

In this section, I've delved into the network subnetting world. Subnetting makes it possible to break down large networks into compact and manageable ones by splitting networks into segments. The details of how this works are beyond the scope of this book, but I've included a subnetting resource in the Appendix if you want to know more.

Step 6: Configure the router's Wi-Fi networks

Setting up a Wi-Fi network is pretty straightforward for most use cases, but it also can be very complicated. Let's walk through the set-up that works for most people.

First, you want to plan the Wi-Fi network name(s) (also known as a Service Set Identifier, or SSID) that you will use. With some routers, you will need a separate Wi-Fi name for each band (2.4 GHz and 5 GHz). I also recommend that you come up with a separate Wi-Fi name for the network your smart devices will connect to (I usually refer to them as Internet of Things, or IoT, devices) and for your guest network. I'll explain why in the security section of this book. Make sure the names are different from the defaults and easy for you to identify.

Next, set up the security for the primary home Wi-Fi network. You will want to select at least WPA2 level security (preferably, WPA3, which is newer and more secure and also backwards compatible with WPA2). Anything less than WPA2 is insecure. If you have a choice between Advanced Encryption Standard (AES) and Transient Key Integrity Protocol (TKIP) encryption, choose AES. TKIP should be used only if you have devices that don't support AES. You probably don't have any of these devices. If you do, you should replace them. TKIP is older, slower, and less secure than AES. Make sure to choose a strong unique password for your Wi-Fi network.

If your router supports band steering, you can use the same network name for both your 2.4 GHz network and your 5 GHz network. Band steering tells Wi-Fi clients to connect to the faster 5 GHz protocols if they support them. However, some clients don't support or work well with band steering. In that case, you will want to set up separate networks with different names.

Next, set up your other Wi-Fi networks, like your IoT and guest networks. If your router supports it, your IoT and guest networks should have limited access to the other devices on your LAN. You also want to give each of these networks different passwords that you don't use anywhere else. Take care in choosing your passwords. They should not be easy to guess, which means they shouldn't contain identifying information in them like your house number, phone number, birthdate, anniversary date, etc. Make them sufficiently long (15 characters is good). Length is a very important deterrent. Remember, you have to enter the password only once per device.

Additional recommended settings to configure:

- Make sure your Wi-Fi network is set to broadcast its name. Setting the name to hidden doesn't stop hackers from seeing the network name but may make it difficult for some of your devices to connect.

- I recommend turning off MAC filtering. MAC filtering restricts which devices can connect to your Wi-Fi network. You have to enter each MAC address of all the devices you want to connect and continue to do this when you want a new device to connect. However, MAC addresses can be copied and impersonated. Some devices change their MAC address for every network they connect to. It also doesn't stop a hacker from snooping on your network. It mostly creates a headache for whoever is administering the network—you!

- Set your channel selection to auto. This means that your router will select the best (most interference-free) channels for your Wi-Fi signals.

- Set your channel width to (1) 20MHz for the 2.4 GHz band and (2) auto for the 5 GHz band. Channel width specifies how large of a pipe is available for data transfer. Wider channels are faster; however, they are more susceptible to interference and more likely to interfere with other devices. Wireless interference is a big concern for the 2.4 GHz band, so I recommend the smaller 20MHz width. It's not as big of a concern for the 5 GHz band.

- Turn on Wi-Fi multimedia (WMM). WMM prioritizes network traffic in a way that maximizes the performance of many network applications like video and voice.

Step 7: Configure the router's security

The most important security step to take with your router is to make sure the firewall is enabled. Next, make sure you disable remote management (the ability to manage your router settings outside of your network). Most people don't need to configure their routers when they are not on their network, and it isn't worth decreasing the security of your router.

I also recommend that you disable Wi-Fi Protected Setup (WPS). WPS simplifies adding wireless devices to your network, usually using either a push button or a personal identification number (PIN). Unfortunately, WPS is insecure and makes it easier for hackers to connect to your network as well. It's not worth the convenience. Just use the traditional password method of adding devices to your network.

We'll discuss more advanced security settings like port forwarding, firewall configurations, and network segmentation in more detail in the security section of this book.

Step 8: Other router settings

- Most routers will have network address translation (NAT) enabled by default. But just in case, check to make sure it is enabled. NAT enables private IP networks to connect to the Internet. If you turn it off, your devices will most likely lose Internet access.

- I also recommend that you disable Universal Plug and Play (UPnP). UPnP is a convenient way of allowing devices on your network to find and communicate with other devices, including devices that aren't on your network. If UPnP is enabled in your router, devices can tell your router to open up ports in your network, which allow access to your internal devices. These can leave your network susceptible to hackers, who scan for those ports and know how to exploit the devices connected to them. Some games and devices won't work properly without UPnP, but you should avoid it if you can, or at least disable it when it isn't needed.

- Quality of service, or QoS, allows your router to prioritize Internet uploads and downloads for certain types of traffic (e.g., video streaming, videoconferencing, and voice over IP) and devices (e.g., computers, smart TVs, and gaming consoles). This helps improve the experience of devices that need solid connections. I recommend taking the time to learn how to enable and configure your router's QoS settings.

Step 9: Back up configuration and set up automatic backups

Once your router is working great, you should back up the configuration. Over time, you may need to play with your router settings, and it's possible you could misconfigure or break something and not remember what your original working settings were. If you have a backup of working settings, you can simply restore your router back to those settings. If your router supports regularly scheduled configuration backups, you should definitely take advantage of them.

Step 10: Test your speed again

Now that you've finished configuring your router, test your speed again. It should still be close to the speeds you were able to achieve in your previous testing. If not, make sure something else on your network isn't hogging the bandwidth and check your QoS settings for any misconfigurations. If your speeds are close to what you were expecting, congratulations! You've configured your router.

Configure Your Mesh Wi-Fi Router

Configuring a mesh Wi-Fi system doesn't deviate much from the steps required to configure a standalone Wi-Fi router. However, there are a few additional steps you may need to take.

Step 1: Connect the mesh system to the Internet and download the mesh system's mobile app

The mesh system, like other routers, will need a connection to the Internet. Also, most mesh systems require a mobile app for management. You'll need to configure an account and an administrator password. Be sure to pick a strong password, and don't forget it! Usually, the app will walk you through the rest of the mesh system set-up.

Step 2: Place the mesh satellites in optimum positions for the best Wi-Fi coverage

This is the most important step for a mesh system. Wi-Fi router placement is always important. Generally speaking, you want to place them high up on a wall, in the open, and away from interfering signals.

You have to go a step further with a mesh Wi-Fi system. Router and satellite placement is paramount for providing strong coverage throughout your home. A general rule of thumb is to place each node halfway between the router and a dead zone but to limit the distance to no more than two rooms (about 30 feet). Each node should be close to a power outlet and as high in the room as possible. Many mesh systems provide guidance and testing capabilities to help you place the satellites in optimum positions.

The last placement consideration is about the devices you want to connect to each satellite. Many mesh satellites come with a wired LAN port that can provide wired devices a strong network connection. You'll need to place the satellite near the wired device (or switch) that you plan on connecting to its LAN port.

I recommend testing your Wi-Fi speeds in various locations in your home to make sure your satellites are placed properly and Wi-Fi coverage is good everywhere.

Step 3: Configure the Wi-Fi, DHCP, security, and other settings

These steps and considerations are the same as for a normal Wi-Fi router, detailed earlier in this book. Follow the same steps for a mesh Wi-Fi router.

Configure Your Managed Switch

Configuring a regular (unmanaged) switch is as easy as it gets. Plug the switch into a power outlet, connect it back to the router using Ethernet, and then plug in LAN devices. No additional configuration is necessary (or even possible). In contrast, a smart (managed) switch takes more steps because it has more capabilities as described in the network devices section of this book (e.g., VLANs, QoS, link aggregation, etc.). What you can configure and how you can configure each smart switch varies greatly across brands and models. This section will cover the more common and popular configurations.

QoS

QoS in a switch works just like it does for routers. It allows your switch to prioritize traffic based on the device plugged in or the type of traffic. If you have a machine (like a video-streaming device) or a type of traffic (like Zoom, Skype, Google Chat, FaceTime, etc.) that you want to make sure isn't interrupted, you can prioritize it using QoS.

Link aggregation

Link aggregation refers to a set of protocols that allows a switch to combine ports to provide failover (keeps working if one network port fails) and bandwidth improvements (combines the capacities of multiple ports into one network connection). A key point to keep in mind is that the device connecting to the switch must support the same link aggregation protocol that your switch is using, or it won't work. Fortunately, there are standards.

Port mirroring

Port mirroring is when traffic from one or more ports on a switch is copied (mirrored) on a different port. Why would you want to do this? Many security appliances work by analyzing all of the traffic on your network to detect intruders and malware. One easy way to send them all the traffic for a switch is to connect them to a port that also mirrors the traffic on all the other ports. This way, it can analyze all the traffic without having to connect to the other devices.

VLANs

VLANs are used to create network segments that are bound by geographic location. Most people have a single LAN that encompasses all of their home network devices. However, some people use VLANs to separate different parts of their networks from each other for security or network organization reasons. If you want to use VLANs, your router and your switches need to support them.

Access Control Lists (ACLs)

ACLs are exactly as they sound. They control which traffic each device has access to. For instance, you can (1) configure only the computers on your network that need access to your router's configuration interface to have access and (2) block all other computers. ACLs can be used in conjunction with, or instead of, VLANs to add security controls to your LAN.

Configure Your Wi-Fi Access Point

Configuring a Wi-Fi access point follows most of the same configuration steps as configuring the Wi-Fi in your router or mesh router system. You want to follow the same steps for configuring a mesh Wi-Fi router but with the following differences:

- **Do not configure DHCP** – Your Wi-Fi access point will not act as your router. In most cases, DHCP should be configured in your router.

- **Make sure your desired access point location is near an Ethernet connection** – Your access point will need a wired connection to a switch or a direct connection to your router.

- **Pay attention to your Wi-Fi channel selection** – Overlapped or duplicate Wi-Fi channels from multiple access points and your router can cause problems with your Wi-Fi signals. This is especially a problem for 2.4 GHz signals, which have fewer channels. Many access points and Wi-Fi routers have an auto setting that should pick the best channels for your environment, but this is something you'll want to check, especially if you are having problems with your Wi-Fi stability.

- **Replicate Wi-Fi network names and settings across access points and your Wi-Fi router** – This helps your Wi-Fi devices seamlessly roam and stay connected to Wi-Fi no matter where in your home they are. If you have a network named MyHomeNetwork on your router, you should also create a network with the same name, configuration, and password on each of your access points. If you make the Wi-Fi network names different, you'll need to configure your existing and new Wi-Fi devices to connect to all the different networks.

- **Look for centralized management software** – Centralized access point management software, like TP-Link's Omada, can control and configure multiple access points and routers from one interface. This makes it easier to configure Wi-Fi settings consistently across the devices.

Configure Your Wi-Fi Extender

In most cases, a mesh Wi-Fi system or Wi-Fi access points are better options for extending the range of your Wi-Fi network than a range extender. They are typically faster. However, there are some cases where using a Wi-Fi range extender is simpler, more convenient, and gets the job done. Here's how to configure a Wi-Fi range extender:

Step 1: Plug in your Wi-Fi range extender close to your Wi-Fi Router

This makes it easier for the range extender to connect to your Wi-Fi router's wireless network so it can extend it.

Step 2: Connect the range extender to your Wi-Fi router's network

Wi-Fi range extenders often use WPS to connect to your router. If your router supports this, you can enable it temporarily to easily connect your range extender without needing to enter a network name or password. Remember, I recommend disabling this feature for security purposes.

If you can't use WPS, you'll need to log in to your range extender and connect to your Wi-Fi router manually using the Wi-Fi network's name and password. Most modern range extenders use an app for configuration. If a range extender doesn't have an app, you may need to look through the documentation for how to connect to the management interface.

Keep in mind, if you have a dual-band router and a range extender, you may need to connect the range extender to both 2.4 GHz and 5 GHz network signals.

Step 3: Name the Wi-Fi range extender's networks

Many range extenders will create their own Wi-Fi network names instead of using the same ones as your Wi-Fi router.

You can leave this as is, but I recommend making the network names the same as your router (with the same password) for a more seamless Wi-Fi experience for your client devices.

Step 4: Unplug the Wi-Fi range extender and move it to its new location

How do you find that new location? Start with moving the range extender to a halfway point between your router and the place where you want to boost the Wi-Fi signal. Then, using the range extender's app or management interface, see how strong its connection with your router is. If the connection is strong, you can probably move it closer to your desired Wi-Fi boost location. If it is weak, you will want to move it closer to your router. Many Wi-Fi range extenders include a feature to help you find an optimal location.

Step 5: Test the boosted Wi-Fi signal

Make sure everything is working. Take a Wi-Fi device to the locations where you didn't get Wi-Fi or had a weak signal. If everything is working well in those locations, then you are done! If not, you may need to play with the location of the Wi-Fi range extender or try a more robust option like a mesh system or Wi-Fi access points.

What I Do: Creating Strong Wi-Fi Throughout My Home

I have prioritized making my wireless network fast and stable. My router doesn't have wireless functionality, so I rely completely on wireless access points. To make sure I have good coverage throughout my home, I have placed a wireless access point on each floor of my home. I'm currently using relatively inexpensive TP-Link access points, which support both 2.4 GHz and 5 GHz (802.11n/ac) signals and provide plenty of speed. I have over 80 devices on my Wi-Fi network, including surveillance cameras and streaming devices, but I haven't noticed any speed or stability issues.

One really nice thing about the TP-Link access points is that I can manage them centrally using free controller software (Omada) provided by TP-Link. This allows me to create and manage the wireless network and connected clients from one spot or by using an app. I can see which devices are connected to which access point and the strength and speed of their connections.

How Do I Do That? Sharing Files Over Your Network

One of the most common uses of a home network is to share files across devices on your network. This is useful for collaborating on files with other members of your household.

Many cloud services like Google Docs/Google Drive, Microsoft OneDrive, and Dropbox have made sharing files with anyone simple. However, another popular use is streaming local videos (movies you've saved to your hard drive, home movies, etc.) and photos to other computers and televisions via smart TVs and streaming devices.

A network attached server (NAS) is the perfect device for storing, backing up, and sharing files across your network. NAS manufacturers like QNAP and Synology make devices that are simple to configure and use. The only downside is that they can be expensive.

You don't need a NAS to simply share files across your network. Here's how to do it.

Windows

1. Open up File Explorer and right-click on the file(s) and/or folder(s) you want to share.

2. Select Give Access to > Specific People. You can choose specific users and their permission levels (whether they can only read the files or read and modify the files), or you can share with everyone. In most home network scenarios, sharing with everyone is often the best choice. Then click Share.

To access the file you shared on another Windows computer, simply look for the name of that computer in the Network section when on another computer. You'll see a list of files and folders previously shared.

MacOS

1. Choose the Apple menu > System Preferences, and then click Sharing.
2. Select the File Sharing checkbox.
3. Click the Add button at the bottom of the Shared Folder list, locate and select each folder that you want to share, select the users you want to share with and their sharing permissions, and then click Add.

To access the file you shared, click on the Go menu and select the Connect to Server... menu item. Then enter the name or the IP address of the computer that contains the shared files and click Connect. You may need to enter the username and password if security was set on the file share. Then, you will see the shared folders.

Home Network Considerations for a New Home Build

Nine years ago, we moved into our current smart home. It was a new construction, but the build was already in progress when we purchased the home. However, I did have an opportunity to specify a few technology items like Ethernet jacks, speaker wire, and outlet locations. Unfortunately, I had only a day's notice, and didn't have enough time to really plan right. All in all, I think I did a decent job.

There are so many things to think about when building a new home. What colors should the house be? What material should the floors be? How should we lay out the kitchen and bathrooms? These take a lot of time to research and decide. Consider putting as much time and thought into the home-networking capabilities of your home. You can make your home network much more robust and flexible if you take a few steps during construction. Here are some things to think about if you are fortunate enough to be involved in a new build.

- Laying networking wire is least expensive during construction. Try two Cat 6 or Cat 7 runs to every room. In this day and age, even the bathrooms should be cabled (you never know). It also may be a good idea to run Cat 6/Cat 7 to a central location on the ceiling or high on a wall on each floor for an access point. If you have a larger house or lots of wireless devices, you may want to consider two locations on a floor.

- Consider runs in locations where you may want to install video surveillance cameras. This includes outdoor cameras. This will allow you to use PoE cameras.

- Think about where you will install smart TVs, projectors, and streaming boxes. Make sure you have Ethernet connections in these locations (and power). These may be in the middle of a wall for a mounted TV or up on a ceiling for a projector. Sure, all these devices could connect via Wi-Fi. But any place you can use wired connections you should. Try to reserve Wi-Fi for truly mobile devices (cell phones, tablets, laptops, etc.) as much as you can.

- Make sure potential ISPs make connections, or create an easy way to make a connection, to your home. At my house, cable (coax) was connected from the street to my house, but fiber was not. However, they did install a couple of underground tubes that ran to my house from the utility pole, which allowed me to install fiber Internet.

- Think about where all the wiring will terminate. I have a small structured wiring panel inside my master closet. All of the Ethernet and coax throughout

my house terminate there. It's a small and incon-
venient spot behind some clothes. When it was first
installed, there wasn't even power, making things
difficult. If I could have decided how that went, I'd
have designed an entire closet (maybe around the
size of a linen closet) that had power and could fit a
server rack. This would make working with the
house networking so much easier.

Key Takeaways

❏ Evaluate how well your current home network and
equipment are functioning before beginning improvements, including the following:

   ❏ How well is your Internet service working?

   ❏ How well is your current networking equipment
working?

   ❏ How robust is your Wi-Fi network?

   ❏ How many and what types of networked devices
do you have in your home?

   ❏ What is working well and what isn't on your
home network?

❏ Create home network goals beforehand to guide your
home network improvements. Think about goals such
as the following:

   ❏ What do you want to make better (speed, stability, security, etc.)?

❏ How many devices do you plan to have on your home network?

❏ Do you want to connect to your home network from remote locations?

❏ Start your home network upgrade by selecting an ISP and networking hardware.

❏ Select an ISP with the upload and download speeds you need for your home.

❏ Consider coverage, security, and budget when selecting a router.

❏ Decide whether you want managed or unmanaged switches.

❏ Determine whether you need wireless access points.

❏ Configure your network, starting with your modem and router.

❏ Learn the key steps to configuring routers, switches, access points, and Wi-Fi range extenders.

❏ Put thought into planning your home network in a new construction build like you would into designing the kitchen and bathrooms.

PART 3

SECURING YOUR
HOME NETWORK

The Importance of Protecting Your Home Network

Private family pictures and photographs. Sensitive financial information. Important software and devices that you and your family count on every day. If someone breaches your home network security, they can compromise all of these things. Bad actors are constantly trying to penetrate and compromise your network, steal your data, and co-opt your computers for malicious intent.

Not a day goes by when hackers aren't probing your network, trying to exploit vulnerabilities. This may sound like hyperbole, but it isn't. The firewall logs for my home router and my business websites show multiple scans and attempted break-ins every day. Luckily, most of them are unsophisticated and easily thwarted. The point, however, remains. Everyone should take their home network security very seriously. A compromised home network can result in identity theft, loss of important data, invasion of privacy and more, costing you time, money, and happiness.

Every home network should be designed and built with security in mind. The good news is that there are plenty of ways to block hackers and protect your home network. In this section, we'll discuss basic, intermediate, and advanced steps you can take to protect your home network. We'll also discuss secure ways to remotely connect to your network.

Basic Network Security Tasks

These basic network security tasks should be the minimum that everyone implements. They can be done easily in ten minutes or less and don't require much technical knowledge. There are no excuses for not implementing these!

Keep Your Router and Other Network Device Firmware up to Date

Your router (and the firewall in it) is your primary defense against hackers penetrating your network. New vulnerabilities are discovered all the time, and you want to make sure that your router has the latest protections. Periodically check to see whether your router has an available firmware update and keep your router up to date. Many routers will notify you of a firmware update. If your router is no longer receiving updates, you should consider replacing it, as it may be too old.

Similarly, you need to keep the firmware up to date on a Wi-Fi access point, a managed switch, Wi-Fi extenders, etc. Cybersecurity experts discover new vulnerabilities frequently, and the best way to protect your network is to keep the devices running your network up to date.

Keep the Client Devices That Connect to Your Network up to Date

Another common way hackers compromise your home network is through security vulnerabilities in the devices on your network. This includes PCs, laptops, mobile phones, smart TVs, surveillance cameras, and more. You can better protect your network by keeping the software and firmware on all of your home network clients up to date as well. I know that everyone hates Windows updates, but they make your home network safer!

Keep Your Antivirus/Firewall Programs up to Date

First, make sure your computers run an antivirus program. While your router firewall is your primary defense for your network, a strong antivirus and firewall program running on your computer is the primary defense for your computer. Most malware comes from sites users interact with, not from hackers getting into your network. Protect the computer from yourself by making sure it has the latest protections your firewall and antivirus program provide.

Make Sure You Have Properly Configured Router Settings

These settings were mentioned in the router configuration section earlier in this book, but they are worth mentioning again. Log in to your router and check the following:

- Remote access (from outside of your network) is disabled. You should restrict the ability to log in and change your router settings outside of your network. You want to be able to configure your router from your home network only (or even from specific computers only, but that's a more advanced topic).
- Your firewall is enabled. This should be the default, but you'll want to make sure!
- You have a strong password for logging into your router. Change the username if you can.

Change Your Home Wi-Fi Password(s)

Please tell me your Wi-Fi password isn't the default (and that you have one and have security turned on for your Wi-Fi network). Implement a strong password for your Wi-Fi network and change it every so often, especially if you have neighbors that are in its range. There's a resource for how to implement strong passwords in the appendix of this book.

Security Emphasis: Always Update the Firmware of New Routers

Most consumers don't realize that many devices bought in stores ship with vulnerable firmware. Even routers you buy on Amazon are subject to this issue. CyberNews researchers recently found numerous flaws in a very popular TP-Link router sold by Amazon[3]. Most of the flaws are fixed in current versions of the router firmware. However, new routers are shipped running old versions of the firmware. Owners of the routers must update the firmware to fix these vulnerabilities; the routers don't update their firmware automatically.

The firmware security vulnerabilities enable hackers to easily compromise the router and gain access to the owner's home network. Here are some of the things hackers can do with control of your router:

- Intercept your web traffic and steal your usernames and passwords to any site, leading to identity theft
- Install malware on the computers in your home network, including ransomware

---

[3] https://cybernews.com/security/amazon-tp-link-router-ships-with-vulnerable-firmware/

- Make your router and the computers in your network part of a botnet group that mines cryptocurrency or attacks another network
- Compromise your video cameras to spy on you

There are more things a hacker can do if they take control of your router. Don't expect your router or other network devices to update their own firmware. Protect yourself by keeping your firmware updated, even in devices you just bought.

Intermediate Network Security Tasks

These intermediate network security tasks take a little more time to implement and even some research. They are more than worth the time for the extra security they provide.

Encrypt Your Drives

This is especially important for mobile devices like laptops, tablets, and mobile phones. If these are lost or stolen, thieves can easily access their drive without knowing the login information for your device by extracting the drive and connecting it to their device. Encrypting your drives makes it much more difficult for thieves to access your sensitive data, even if they have physical access to your device. If encrypting your entire hard drive seems too drastic, or like too much work, you can encrypt just the sensitive files you want to protect.

Set Up a Guest Network

If you don't have a guest network, you may find yourself in a situation where you have guests over and you have to give them your password so they can access your network. This gives your guests not only your password, which may be sensitive and used for different accounts (you should not re-use passwords), but also access to all of your networked de-vices (printers, network shares, PCs, etc.). Sure, you probably trust the people who have your password, but do you completely trust that all the software and apps they use are free of malware that can compromise the devices on your network? You shouldn't. A proper guest network al-lows you to set a separate and shareable password for your guests. It also protects your private computer resources from being compromised by guests. Many routers and ac-cess points make configuring a guest network simple. If yours does, it's worth learning how to configure it. If not, consider purchasing a router or access point that has the feature.

Set Up Regular Backups

Although not a direct protection against a network breach or computer infection, having a backup strategy can save you from catastrophic loss of data if there is a breach, or even if hard drives or computers fail. It also can help you restore your systems in case they are exposed to data corruption or ransomware. Here are a few options for backing up your systems.

Local backups

The easiest and quickest strategy to implement is setting up a local backup somewhere in your house. This can be as simple as occasionally copying all of your important files to a separate directory or hard drive on your computer. A step up would be to buy an external hard drive to store a backup on. One more step up would be to back up your files to a network attached storage (NAS) machine like the ones from Synology or QNAP. With some spare hard drives and low-powered computers, you can also pretty easily build your own NAS or turn your desktop PC into a NAS.

You can combine different types of local backups for better reliability and redundancy. For example, you could:

- Store all documents you want to back up on a NAS. Then you can use an external hard disk to back up the NAS. Some NAS devices have ports for external hard disks designed just for this scenario and can automate the backups.
- Back up files to a separate directory/disk on your computer and then regularly create a secondary backup on an external hard drive. This gives you backup redundancy.

If you are going to use a local backup strategy, I highly recommend that you use a NAS, an external hard drive, or a drive on a computer different from the computer that contains the files you are backing up. Having multiple copies on one computer as a backup is much riskier.

Back up documents to the cloud

Cloud storage for consumers and businesses has become a very hot and competitive business. Many services offer free storage, such as OneDrive, Dropbox, and Google Drive. These services allow you to pay for additional storage. You can install apps on your computers and mobile devices to automatically back up important files to these cloud accounts. You can also create and directly access your documents in the cloud. All of these services have your data backed up and replicated in multiple locations to keep your data safe. If you want to back up many photos or large video files, these services can become expensive.

Remote backup

OneDrive, Dropbox, Google Drive, and similar services are more geared towards syncing your data than backing it up. Their primary goal is to keep your files synced across multiple machines and devices, thereby providing access anywhere. In contrast, remote backup systems are best if you're concerned about backing up large amounts of data and want a higher level of security. Cloud backup providers encrypt your files before storing them, which makes your data more secure because the provider doesn't have the key to decrypt your data. Popular cloud backup services include Backblaze, Spideroak, and IDrive. All of these solutions come with software you can install on your PC(s), and schedule automated backups so you don't even have to think about it. Some of these solutions have apps that will back up your mobile devices too!

Alternatively, you could set up a remote server at a friend's house for remote backup, but that may be more work than it is worth!

Combining backup strategies

The more copies of your data, the less risk of data loss. A popular backup technique is the 3-2-1 backup strategy which stands for the following:

- 3 copies of your data
- 2 local copies on two different devices (original and backup)
- 1 offsite backup

This means combining the first strategy in this section (local backups) with one of the other strategies (backing up documents in the cloud or remote backup). For example, you could have a process that backs up to your NAS frequently and then performs a daily remote backup. Some cloud backup services have direct integration with popular NAS providers to make this process even easier.

Make sure you automate whatever strategy or strategies you decide to implement. Reliable backups are critical and having to remember to back up a file is unreliable. There are software and apps available to automate backing up locally, with a cloud syncing service, or remotely. Also, occasionally restore backups to ensure that your backup will work when you need it to.

What I Do: Backing Up My Important Files

I employ the 3-2-1 strategy. I have a NAS, and each night I back up all important documents, photos, videos, and other important files to a separate machine from my NAS. SpiderOak is the cloud backup service I use to remotely back up my NAS. All computers in my house have access to the NAS. My family knows their important files are backed up locally and remotely every night.

One added benefit of using SpiderOak is that they keep copies of previous versions of files. If I haven't kept an old version of a file, I can find it in my cloud backup. This doesn't take up much extra storage space because of SpiderOak's deduplication (they store the changes to a file instead of a new backup each time). They also have a mobile app and web interface that allow me to access my backup files from anywhere.

Advanced Network Security Tasks

These tasks may take a significant amount of research and planning. Many of these concepts go beyond the basic use of routers and antivirus software. Take the time to learn how to implement these tasks to take your home network security to the next level.

Network Segmentation

If your IoT devices are running on the same Wi-Fi network as all of your other devices, they probably have access to the rest of your network. Unless you've done something explicitly to limit their access (e.g., VLAN, access control lists, a separate IoT Wi-Fi network), they are probably members of your network like any other device.

You may be asking why IoT devices need to be segregated from the rest of the network. The answer is that IoT devices have been hacked en masse. Lightbulbs, switches, thermostats, refrigerators, network cameras and many more IoT devices have been hacked. Some have been compromised to attack other people and companies; some have been used to attack the device owners. IoT devices have revealed owner Wi-Fi passwords.

IoT devices are not known for having the best security features, and they are usually always connected and accessible from the Internet. Once these devices are compromised, they can be used to probe and attack other devices on your network. Sensitive personal information stored on your computer could be compromised. This is why you want to protect the rest of your network from these devices.

Segregating your IoT devices can be a complicated affair. Some ways of doing it involve modifying routing tables and setting up separate subnets. But nowadays, newer routers and access points have features where you can set up a network just for these types of devices, which secures your other network devices from them. This works very similarly to a guest network. You just end up putting your IoT devices on a separate network. When you are looking for a new router or AP, look for this function. Yes, we live in a world where we need to protect ourselves from lightbulbs and thermostats!

Monitoring for Unknown and New Devices

Do you know all the devices that should be on your network? Do you occasionally check your network to see whether unknown devices have connected? Ideally, your router allows you to monitor the devices connected to your network. To have your router alert you when new devices connect to your network would be even better. Most routers don't have that function, but you can use software that scans your network for new devices and sends alerts. There are free and paid options for whatever operating system you run.

Centralize Your Logs and Monitor for Breaches

You can do a lot more than monitoring what devices are on your network. Most people have many devices that interact with their home network. Some devices, like personal computers, routers, switches, and access points, log their activities. By examining these logs, you can detect all kinds of security breaches. However, all these devices produce a lot of logs to monitor. Going to each machine and application and inspecting the logs would take a lot of time. Fortunately, you can put log monitoring for security breaches on autopilot by centralizing your logs and using a log-analyzing tool—e.g., Graylog (which I use), Eliasticsearch, ELK, and Splunk. What can you do and monitor with these tools?

Failed login detection

Windows event logs and syslogs contain records of authentication failures (such as entering a bad password). With centralized logging, you can set up an alert that emails you whenever there are too many (three is too many for me) unsuccessful authentication attempts in a minute. Authentication failures can indicate that someone is trying to break into your computer. So alerts for these events are a good line of defense.

Notification of new machines on your network

If you assign all devices on your network an IP address, as I recommended in the router configuration section of this book, then you can be notified whenever an IP you haven't mapped joins your network. This way you know when anyone new connects to your network.

Notification of users connecting to your guest Wi-Fi

Users who connect to guest W-Fi aren't usually already mapped to an IP. Your access point or router probably logs when users connect to your Wi-Fi networks, and you can set up your log analyzer to alert you whenever someone connects to your guest Wi-Fi. This is to make sure someone isn't mooching off your Wi-Fi unexpectedly.

Ability to search for particular events

Sometimes, things go wonky on a network, or things may happen and you'll have no idea why. Often, you can successfully troubleshoot by looking through the logs. You can search through logs to help understand why applications have crashed, PCs rebooted, and to investigate unusual traffic in your network.

Set up a dashboard with key security metrics

You can set up a security dashboard that quickly allows you to monitor key security metrics. For example, I have a dashboard that shows me the number of times someone logs into each of the machines and a graph of login failures. All of this information is filterable by IP address and presented in charts that I can drill down into. For kicks, I also keep a histogram of how many security-related messages are in all of the aggregated logs. It's a ton!

Malware Prevention at Network Level

You can process incoming Internet traffic for your home network and block malware and known hacker and hacking attempts. This is another feature that some higher-end routers and router software have, like pfSense. They are usually referred to as an intrusion protection system (IPS) or intrusion detection system (IDS). The easiest way to do this is to buy a router that has these features or a dedicated firewall security device.

Have a Disaster Recovery Plan

One often-overlooked part of home network security is having the ability to recover from a network failure or a hack that compromises your network. You can take all the steps above and still get hacked. There are no foolproof solutions, just ways to make hacks less likely. For this reason, you want to be able to recover from a hack. A disaster recovery plan will help get your home network back up and running smoothly. A good network disaster recovery plan includes the following:

1. **The ability to detect outages or other disaster effects as quickly as possible.**

   As discussed in the Advanced Network Security Tasks section, monitoring your network is key to detect breaches and outages as soon as possible. The sooner these things are detected, the less damage can be done to your network and the quicker you can recover.

2. **Alerts sent to the network administrator (probably you) so action can be taken.** Timely notifications of network problems are important as also discussed in the Advanced Network Security Tasks section. Again, the person responsible for maintaining the network should find out about problems

as soon as possible to thwart and recover from threats.

3. **Documented steps to isolate affected systems so that damage can't spread.**

    Often, the first step in managing a network breach is to isolate the affected parts of the networks so they can't infect or impact other parts of the network to limit damage.

4. **Documented steps to repair and/or rebuild the affected systems so full home network function can resume.**

    Once you've diagnosed and stopped the threat, the repair work can begin. Having the recovery steps documented makes this process easier.

A hallmark of good disaster recovery planning is documentation. Sometimes it's hard to think clearly in a crisis, and being able to follow well-written documentation is critical to remaining calm. Good disaster documentation includes the following:

- Well-documented configurations of key network devices such as your router, managed switches, file servers, and access points.

- A network diagram of your home, which includes network device locations, IP addresses, hostnames, the way they are connected to your network, VLANs, etc.

- Any manual configuration steps needed before and after restoring configuration files to your key network devices.
- Locations of key configuration files and backup files, as well as the way to gain access to them.

What I Do: Securing My Home Network

My first line of defense is my pfSense firewall and router. pfSense is actively maintained, and I make sure I keep up to date with new firmware. I've made certain that remote access to my router is disabled. I've also taken additional steps to ensure only a couple of computers on my LAN can access my router's user interface, for additional security.

Snort and pfBlocker are packages available with pfSense, which I use to further protect my network from bad actors. Snort detects known intrusion signatures—telltale signs that show a system is being attacked—and blocks hackers trying to access my network. pfBlocker takes a different approach by reading regularly updated lists of known computers of hackers and other bad actors and blocking them from accessing my network.

I have more computers in my house than most, because I repurpose old hardware into file servers, multimedia distribution systems, virtualization hosts, etc. Every single one of these machines runs a firewall with rules to block other devices in my house from accessing popular network services (e.g., web connections, file and printer sharing, remote desktop, etc.). These rules are customized for each machine. My file server obviously allows most computers to access its files, and my web servers allow HTTP/HTTPS connections. By default, however, my firewall rules are set up to block all my IoT devices from accessing anything unless I write in an exception. All of my computers run Linux, and I use the UFW firewall software to accomplish this.

For additional Wi-Fi security, my home is separated into three Wi-Fi networks: home, guest, and IoT. Each Wi-Fi network has a unique identifying name. Computers on my home network have access to all other computers on my network, and security is a bit laxer. Computers on my guest network cannot access any other computer on my network and are also restricted from hogging all of the bandwidth on my Wi-Fi. Computers on my IoT network are restricted from accessing most computers on my home network, with some exceptions. For instance, my security cameras need to save their video to my surveillance system hard drive, so they have access to that machine. My home automation controller (Home Assistant) needs to control many of my IoT devices, so they are permitted to access each other.

I put in an additional restriction for my security cameras. I use firewall rules in pfSense to block them from accessing the Internet. This way, hackers cannot look at the video from my cameras, and my cameras can send video not to the Internet but to my home network only. I can view my cameras away from home by connecting to my home network VPN (more on that later).

Graylog is the software package I use to centralize all of the logs on my network. I have all the system logs from each computer, my router, and my access points sent to my Graylog service for processing and storage. Graylog is set up to alert me when unknown computers join my network (wired or wireless), when there are too many failed login attempts to any machine, and even when there are power outages (it receives information from my uninterruptible power supply). Additionally, from one interface I can search through the logs of all of my machines when I am trying to hunt down a problem or an intrusion.

How Do I Do That? Perform a Security Audit on Your Home Network

Security audits are a regular part of doing business for IT departments in companies. Some industries have regulations requiring them to perform periodic security audits. And yet the same is not true at home, of course. Most people never actually test the security of their home network. Here are a few things you should do at least once a year to audit your home network:

- **Run a network scan to determine all the devices connected to your network.** Make sure you recognize all the devices on your network. If you log and set up alerts as to when new devices join your network, you'll always know and won't need to do this scan. Nmap is my favorite tool for doing these types of scans, and there are many Nmap-based tools freely available for Linux, Windows, and MacOS.

- **Review your firewall and port forwarding settings on your router.** These settings greatly affect how people outside your network can access your network. Make sure you know why all the firewall rules are there and what they do. Review your port-mapping settings and make sure you still need them. The less entries you allow into your network, the more secure it is.

- **Test for rogue Wi-Fi access points.** A rogue access point is an access point that allows for a Wi-Fi connection to your network, which you didn't install or set up. You'd probably find one of these when running a network scan. Additionally, some wireless routers and access points have the ability to scan your network for rogue access points.

- **Perform a penetration test.** The only real way to know for sure whether your home network is secure is to see how it holds up to attempts to compromise it. Penetration tests (often referred to as "pen testing") use tools that can simulate standard attacks on your network to see how well your protections hold up. IT professionals often contract with a security firm to perform pen testing, but the costs are usually higher than home network administrators would pay. Fortunately, there are tools out there that you can learn on your own to start doing pen testing and vulnerability scanning, including Metasploit, Wireshark, Hashcat, and Hydra. These tools take some time to learn, but they are worth it.

Remotely Connecting to Your Home Network Securely

Accessing the resources on your home network remotely can be convenient. Maybe you want to access a file that is only on one of your computer's hard drives. Maybe you want to see what's on a network camera, for security purposes.

One way to do this is to set up a secure remote desktop application. These run on a computer in your home and allow you to view and control that computer's desktop from another computer outside of your home network. You can control that computer similarly to as if you were sitting in front of it. Some popular options for remote desktop software are Teamviewer, Chrome Remote Desktop, Microsoft Remote Desktop, GoToMyPC, UltraVNC, and LogMeIn.

Some of these are free and/or have free versions. If you decide to use one of these, make sure you pay attention to the ways of setting it up securely. You don't want to leave a low security connection to a desktop on your network.

The biggest downside of remote desktop options is that you have to leave a desktop running so that you can connect to it. Also, that desktop should not be in use by anyone else when you connect to it remotely. Otherwise, you'll have two people trying to control the same desktop at the same time. These are a couple of reasons why I prefer to set up a virtual private network (VPN) connection. A VPN doesn't require the use of a desktop environment on a computer on your home network. It works by giving any device you are using, including laptops and mobile phones, direct access to your network. Essentially, it makes your device work as if it is on your network from anywhere in the world.

Many routers come with VPN capabilities and offer a simple set-up. Another simple way of setting up a VPN is to buy a security appliance that does VPN connections. I mentioned some examples of these in the other network devices section of this book. Also, if you are technically inclined, you can install and run your own VPN server on a computer in your network.

There are many different VPN protocols. They all have differences, including varying levels of security. The ones I recommend you consider are IPSec and OpenVPN. They both combine security with flexibility and wide device support.

What I Do: OpenVPN via pfSense

OpenVPN was a straightforward VPN choice for me, because I use pfSense as my router, and pfSense supports OpenVPN. I use OpenVPN connections securely on my mobile phone to control my smart home remotely, view security cameras, and sometimes to print. I also use OpenVPN on my laptop to view internal resources like some of my websites, videos, virtual machine hosts, and to securely connect to machines at my leisure. It works consistently, is plenty fast, and has become a feature I take for granted.'

Key Takeaways

❑ Implement basic security steps, including the following:

> ❑ Keep your router and other network device firmware up to date.

> ❑ Keep your client devices (mobile phones, laptops, PCs, etc.) up to date.

> ❑ Make sure you keep your antivirus/malware scanner software up to date.

> ❑ Make sure you've properly configured basic router security settings.

> ❑ Make sure you have strong home Wi-Fi passwords.

❏ Implement intermediate security steps, including the following:

    ❏ Encrypt your PC and mobile device hard drives.

    ❏ Set up a guest Wi-Fi record.

    ❏ Set up regular backups of important files.

❏ Implement advanced security steps, including the following:

    ❏ Segment the various parts of your network from other parts (e.g., IoT, guest devices, surveillance cameras, etc.).

    ❏ Monitor your home network for unknown devices.

    ❏ Centralize your computer logs and monitor for security and failure events.

❏ Create a disaster recovery plan for your network.

❏ You can securely connect to your home network remotely by using remote desktop software and/or a VPN.

PART 4

TROUBLESHOOTING AND MAINTAINING YOUR NETWORK

Keep Your Network Running Smoothly

Most of the time your home network is a reliable afterthought when you set it up correctly from the start. However, even a properly set up network will take some maintenance to keep it running well. Also, sometimes problems are introduced into your home network from adding devices, bad actors, and neighbors adding devices without asking you first. In this section, we'll discuss (1) ways to troubleshoot and fix common home network problems and (2) basic maintenance steps everyone should take.

A Computer on My Network Can't Access File Shares and Network Printers

Few things will frustrate users of your home network more than not being able to print when they want to print, and not being able to access important files when they need to. Here are a few things to try that will help you diagnose and troubleshoot the problem.

- **Can other computers on your network access the file share or the printer?** If so, the problem is most likely with your computer, not the printer or the file share. If no computers can access the printer or the file share, double-check that the printer and/or the computer with the file share are properly configured.

- **See whether you can ping the printer or the computer that you can't access from the computer having problems.** You'll need to know the IP address of the computer/printer you want to ping. You can use the ping command by accessing the command line/terminal on Linux, MacOS, and Windows. If you can't ping them, you'll want to double-check that they are properly connected to the same network as the computer you are using.

- **Your computer's firewall could be blocking connections to and/or from the other computer or file share.** Check your computer's firewall settings to make sure printer and file sharing connections are allowed. You should also check the firewall/security settings on the printer and computer hosting the files. For example, if you are on a Windows computer, make sure "Network Discovery" and "File and Printer Sharing" are enabled.

- **Check that you are using the right credentials to connect to the file share.** If the file share is set up with security, you'll need to know a username and password that can access the files.

Your Computer Can't Access the Internet

Here are some troubleshooting steps if you find yourself in a situation where your Internet connection isn't working:

- **Check to see whether other computers can access the Internet.** If they can't, try rebooting your modem and your router. If after the reboots, no devices have Internet access, you should call your ISP and check to see whether there is an outage or they have limited or disconnected your Internet connection for some reason.

- **If just one computer on your network can't access the Internet, it is likely a hardware issue.** If you are using a wired connection, replace your network cable. If you're using a wireless connection, check to see that your Wi-Fi is connected.

- Go into your computer settings and make sure your network connections are enabled.

- **Disable your firewall temporarily.** It could be malfunctioning. I recommend doing this only temporarily to see whether you regain your Internet connection. If disabling your firewall works, you will need to examine your firewall settings to find the problem.

- Scan your computer for malware.

Your Computer's Network Speeds Are Slow

Trying to browse the web when pages are loading slowly, and waiting a long time for files on your home file shares to open can be really frustrating experiences. These can both be caused by a slow home network. Troubleshooting the speed on your home network can get complicated. Here are a few ways to assess and fix the problems:

- First, make sure you understand what is slow. If you are accessing a website and it is slow, it could be the website having problems, not your computer. Are other websites fast? If so, it is probably not a problem with your network.

- Are file transfers slower than normal? Try accessing those same resources from another computer in your home. Are they still slow? If not, then the problem is probably your computer and/or its connection to your network.

- If all your computers are experiencing slower Internet speeds than normal, try rebooting your router and modem. Sometimes these devices need to be restarted to function properly.

- If your Internet speeds are slow, check to see what other people are doing on your network. Depending on the bandwidth of your Internet plan, things like streaming 4K movies and downloading tor-rents could slow down your Internet connection.

Your problem could also be your ISP or your Wi-Fi connection. I'll discuss in detail how to address those issues in the next session.

Fixing Common Wi-Fi Problems

Have you noticed how important the Wi-Fi connection in your home has become? With the increase of mobile devices and other wireless devices, your home Wi-Fi is more important than ever. Cell phones, streaming devices, smart bulbs, home security systems, washers and dryers, smart speakers, TVs, and so many other gadgets now require Wi-Fi connections to access all the geeked-out features that caused you to buy them in the first place.

The significant reliance on your Wi-Fi means it's important to make it fast, secure, and stable. We will discuss the three major problems most people have with their home Wi-Fi (some of which you may not even know you have) and, more importantly, detail ways to diagnose and fix these problems.

Diagnosing and Fixing Slow Wi-Fi Speeds

Is your Wi-Fi slow? Do you sometimes find yourself waiting a long time for your app to load, or an attachment to download, or for a web page to come up on your mobile phone? Do your favorite streamed shows buffer and/or stream at lower than HD quality? The root cause of your Wi-Fi slowdown can be hard to find, but don't worry. I'll show you multiple ways to diagnose and fix Wi-Fi speed problems.

Issue #1: Your Internet connection is really slow

Symptoms:

- You don't pay much for your Internet service.
- Your Internet connection is really slow when multiple people are using it.
- You haven't properly configured your router's QoS settings.

Diagnosis:

First, find out what bandwidth you have. See the Evaluate Your Home Network section in part 2 of this book to make sure you have enough Internet bandwidth. If you determine your bandwidth is fine, then take a look at your QoS settings. These settings can determine which types of Internet traffic and which devices have priority so that you can use your Internet bandwidth efficiently. For instance, you may want to prioritize video chats and streaming video over torrent downloads and web surfing to eliminate video buffering pauses. These are the types of scenarios you can configure with your router's QOS settings.

After walking through the steps to test your Internet speed and estimate your needs, do you find you have a fast-enough Internet connection? If not, this could be a major reason your Wi-Fi seems slow. You must have enough bandwidth for all the devices (not just wireless) that access the Internet. If you don't, it's time to upgrade your Internet connection. You also want to make sure you have modified your router's QOS settings to prioritize the Internet traffic that is most important to you.

Issue #2: Other devices are hogging your Wi-Fi bandwidth

Symptoms:

- There is a lot of Wi-Fi traffic going across your LAN.
- Somebody is on your Wi-Fi network, and you don't know about it.

Diagnosis:

Do you have a NAS device? If you do, chances are you have backups going across your network. You may stream music, pictures, and videos wirelessly on your network, using something like Kodi or Plex. These will hog a lot of bandwidth on your Wi-Fi.

Do you run some type of networked home surveillance system, or do you have some wireless cameras in your home? Having these on your network can take up more bandwidth than streaming Netflix.

Have you run security checks and malware scans on the devices on your network? (See part 3 of this book for guidance.) If not, you could have malware that is sniffing across your network, doing nefarious things. In addition to compromising your devices and data, this can slow down your network!

Can you identify every device that is connected to your network? You want to make sure you don't have your neighbors connecting and using up your bandwidth.

Make sure you know what is going on in your network and what the impacts may be. If you have items on your network constantly sending traffic (like video cameras), try to make them wired devices instead of wireless. Wireless may be convenient; however, if there is an option for a wired connection, you should take it and reserve the wireless bandwidth for mobile devices and others that are hard to give wired connections to.

Go to your router and/or access point settings and see whether you can tell which devices are connected to your wireless network and what bandwidth they are using. Many routers and access points have traffic charts to detail network activity.

Issue #3: Some of your wireless devices are old

Symptoms:

- You still have 802.11g devices on your network.
- Your devices don't support the 5 GHz protocols.

Diagnosis:

Which protocols do your mobile devices support? If you have a lot of devices that support only 2.4 GHz protocols (802.11g and 802.11n 2.4 GHz), your speeds are limited and your Wi-Fi is subject to more interference than the 5 GHz protocols. Moreover, you may be forced into running your wireless router/AP in a mode that lowers the Wi-Fi speed of all devices that connect to it just to support the older devices.

What protocols does your wireless router/AP support? Does it have multiple radios capable of separating the protocols so that it doesn't need to run in a "mixed mode" that slows down speed for the faster devices?

How long has it been since you upgraded your router/AP? Your router may be too old to support the newer protocols your mobile devices are capable of using. The age of your router/AP also has security implications (more on that later).

If you have a lot of older devices on your network, and you have to support older wireless protocols, you are probably slowing down your network speeds. You may consider upgrading your router/AP to one that can support both older and newer protocols without slowing down the faster devices and/or upgrading your older Wi-Fi devices. Make sure to take the wireless protocol support of a device into consideration when making a purchasing decision, both for your router and your client devices. For example, if you're purchasing a laptop in 2022, look for one that supports at least Wi-Fi 5 (802.11ac) and preferably Wi-Fi 6 (802.11ax). Remember that even if you have a client device that receives data at high speed, it can do so only if your Wi-Fi setup supports it.

Issue #4: Your Wi-Fi signal is weak

This is a common cause of slow Wi-Fi speeds. We will talk more about signal strength and stability later in this section.

Fixing Common Wi-Fi Security Problems

One of the easiest ways to hack into a home network is to attack through its Wi-Fi network. Have you taken steps to secure your Wi-Fi? Many pages of text could be devoted to ways of doing this. In this book, I'll focus on the actionable and important ones that will keep you covered against most vulnerabilities.

Issue #1: Your router is too old

Symptoms:

- New firmware for your router hasn't been released in years.
- Your router doesn't support dual-band wireless.
- You don't get the Internet speeds you think you should get from your ISP.
- You have to reboot your router frequently to keep it working properly.

Diagnosis:

Is your router more than five years old? Check to see what the latest firmware release is. If it has been a few years, your router may not be able to provide adequate and up-to-date security.

Does your router support 5 GHz protocols like wireless N and AC? If it doesn't, it may be too old. Did you run the speed tests mentioned earlier in this book? Were your speeds significantly lower than the package you bought from your ISP? There could be a few reasons for this, but one reason may be that your router is too old to provide the speed that you are paying for. Do you have to reboot your router regularly to keep it functioning?

If your router is running on old software, it won't have the latest security protections. Make sure and check for newer firmware for your router. If your router is outdated, consider upgrading it to properly protect your network and possibly give you faster Wi-Fi speeds and more capabilities.

Issue #2: You haven't changed the default settings on your router

Symptoms:

- Your router login username and password are something like "admin" and "password."
- The name of your Wi-Fi network (SSID) includes the brand of the router.
- Your router's firewall isn't enabled.
- You are not using WPA2 or WPA3 security.
- Your router admin interface is accessible on the Internet.

Diagnosis:

Log in to your router. (Is the username/password "admin"/"password"?) Check to make sure that your firewall is enabled, that you are using WPA2 and/or WPA3 security on your Wi-Fi, and that you have a non-default name for your Wi-Fi network. Find the setting to disable remote access (access from the Internet) to your router.

There are many more steps to properly secure your router; however, if you take care of these, your Wi-Fi network will be much more secure. Leaving settings at the defaults allows hackers to more easily compromise your home network. Many hackers target specific brands of routers. Leaving the brand in your Wi-Fi name signals to hackers the brand of your router. It also indicates that it may have other default settings they can exploit.

Issue #3: You don't have a guest network

Symptoms and diagnosis:

The symptoms and diagnosis here are simple. You never set up a guest Wi-Fi network, and/or your router/AP doesn't have that functionality.

As mentioned earlier in this book, if you don't have a guest network, you may find yourself in a situation where you have guests over, and you have to give them your password so they can access your network. This gives guests access to your entire network, and you may be giving them a password you use for other services. Be sure to use a router or access point that has a guest network feature to protect your network.

Issue #4: You have not segregated your IoT devices from the rest of your network

Symptoms and diagnosis:

The symptoms and diagnosis are simple here as well. We discussed network segmentation earlier in this book, so I won't repeat the details here. In short, you want your IoT devices to run on a network segment that is different from all of your other devices to decrease the chances that a hacked device can impact the rest of your network.

Fixing Wi-Fi Stability Problems in Your Home

We've all experienced Wi-Fi signal cutting out. We're also familiar with weak and dead Wi-Fi zones that cause our streaming videos to buffer, or perhaps kick us out of a mobile game. But with a little planning, your Wi-Fi signal can be strong throughout your home. Here is how to identify and fix Wi-Fi stability problems.

Issue #1: Your router is your wireless access point

Symptoms and diagnosis:

Most routers also have wireless AP capabilities. This can be a cost-effective way, especially for smaller homes that need only one wireless AP, to provide coverage for the entire house. However, the lack of flexibility with where you can place the router can be a problem. For most homes, the ideal place to put a single wireless AP is somewhere near the middle of your home high up on a wall or the ceiling. This spot is usually not an option because routers need to be near the modem provided by the ISP, which is often in a room nowhere near the middle of the house. Also, many people have a modem provided by the ISP that also functions as the router and AP, which further constrains its placement.

If you can't place the router in a good spot in your home, or you have a larger home that a single router can't cover, you will have spots in your home that aren't covered or have poor wireless performance. You can fix this by buying a Wi-Fi range extender or a wireless access point, or by replacing your router with a mesh system. For more information on which of these options are best for you, refer to part 2 of this book where the details of these devices are discussed.

Issue #2: You have too many Wi-Fi devices on your network

Symptoms and diagnosis:

Diagnosing whether you have too many Wi-Fi devices on your network can be difficult. The issue I am diagnosing here is different from the one when too many devices hog your LAN or Internet bandwidth, which I wrote about earlier. Theoretically, most wireless access points can have 255 devices connected at a time, but that doesn't mean they'll work well. You probably won't find the maximum limit in your access point's documentation. If you have only one access point, I'd start to worry about having too many Wi-Fi devices when you get to twenty or so devices connected at a time. At this point, depending on how active those devices are, you may also start to run into bandwidth issues. I mentioned how to solve the bandwidth issue earlier. If you believe you have fixed your bandwidth issues, and you are still experiencing slowness, you may want to upgrade your access point or add an access point to distribute the load of your wireless devices.

Some of you are thinking that you'll never get to twenty or more wireless devices. But it can happen faster than you imagine. Smartphones, tablets, streaming sticks, video game consoles, smart TVs, smart hubs, smart bulbs, thermostats, smart plugs, smart refrigerators, smart toothbrushes, smart scales, smartwatches, smart smoke detectors, smart washers, and smart dryers... I think you get my point. More and more Wi-Fi–enabled devices are entering your home.

Issue #3: You aren't using the best Wi-Fi protocols

Symptoms:

- You have only 2.4 GHz wireless.
- You don't have wireless AC.
- You have Wi-Fi 6, wireless AC, or 5 GHz wireless N, but your client devices don't use them.

Diagnosis:

Go check the wireless settings on your router. Do you have 5 GHz N available? Do you have wireless AC available? Wi-Fi 6? Are any of your devices connecting to the 5 GHz protocols?

2.4 GHz Wi-Fi (both G and N) can be very unstable in today's world. Those protocols deal with interference from other devices (like microwaves, phones, security systems, and baby monitors) and neighbors' running 2.4 GHz wireless networks. This interference can cause speed slowdowns and even drops in your Wi-Fi connection. The 5 GHz protocols are much more resistant to interference. They are also faster. The downside is that they don't cover as much area as the 2.4 GHz protocols, and fewer clients support them. My suggestion would be to avoid buying devices that don't support 5 GHz or newer protocols if possible.

Where to Get Home Network Help

It would be great if all of our home network builds and improvements went off without a hitch! But that's not reality. Sometimes, you have a hard time getting a device to work as you want it to. Other times, something that was working fine for a while suddenly stops working. Also, you may be looking to assist others, like a friend you helped to build a home network.

When these things happen, the "yourself" part of DIY home networks can be a frustrating limitation. The good news is that you aren't actually alone. Home network help can come from many places.

Home Network Device Manufacturers

The company that made the home network device or system you are struggling with may have online support. Many manufacturers provide manuals, knowledge bases, and firmware updates directly to consumers from their websites. Often, the problems with some of the devices I've used have been solved by simply upgrading the firmware. Look for a technical support email or ticketing system. Some companies even have online chat features where you can get immediate help.

Home Network Device Support Forums

Another support channel you often find on manufacturer's websites is a support forum. It's common practice for manufacturers to employ a few moderators and run a forum for customers to discuss and help each other with their products. Other times, product enthusiasts band together to create, run, and support a community-based forum. Since they have no official affiliation with the manufacturer, you probably won't find official help, but unofficial help can be really good. Official and unofficial forums can provide excellent home network support.

Large Support Organizations

So far, all the options I have talked about have been free, but there are good pay options too. Large nationwide chains like Best Buy's Geek Squad and HelloTech can set up, configure, and troubleshoot home network equipment. I personally have used Geek Squad to help set up A/V equipment I bought at Best Buy, and it was a good experience. They also work with equipment you purchase at other places. Both have minimum charges, so it may be best to have them help you with a few things on a single trip.

Local Home Network Help Companies

Depending on where you live, you may be able to find a smaller local company willing to install, configure, and troubleshoot home network issues. In fact, many of the technicians that work for places like Geek Squad and HelloTech are associated with a smaller, local home network help company. You just have to be able to find them. Look at places like Yelp. Searching for computer repair shops and electricians may yield some leads in addition to searching for home network companies.

Social Media

Although social media is well known for misinformation, it also contains a lot of useful advice information. YouTube is filled with a plethora of home network installation, trouble-shooting, and how-to videos. Facebook has private and public groups dedicated to many home network topics and products. Companies often create YouTube channels and Facebook groups with tons of helpful information. Some even use a Facebook group as their official support channel.

There's also Twitter. Most companies have an official Twitter account. Twitter usually works better for communicating about service status and new products. But you can also tweet about the problems you are having with the company's products. I've done this a few times and gotten good responses, and sometimes even help in my direct messages. You can also post issues you are having with popular relevant hashtags, and sometimes the community will help you and/or point you in the right direction.

Home Network Help Blogs and Websites

There is a ton of information about home networks on the web. You'll find some of my favorite sites in the Appendix. These sites are filled with product reviews, how-tos, and other home network help from real people who have done the same things you are trying to do. By the way, I've got a useful home network site as well—HomeTechHacker.com[4].

Maintaining Your Home Network

Benjamin Franklin said, "An ounce of prevention is worth a pound of cure." It's better to prevent problems than to have to solve them later. Regular maintenance of your network allows you to plan for needed changes and prevent problems. Here are some steps you can take to maintain the speed, reliability and security of your home network:

- Keep the firmware on your router, access point, switch, computers, mobile devices, printers, and all devices on your home network up to date.
- Change key passwords to your network devices from time to time (try at least once a year).
- Test your Internet bandwidth from time to time to be sure your ISP is giving you what you are paying for.
- Perform annual penetration testing of your network (see part 3 of this book for more details). Check for and plug vulnerabilities in your home network.
- Check your Wi-Fi signal strength and stability from time to time. Changes in your environment (new

---

[4] https://hometechhacker.com

Wi-Fi devices and devices that interfere with your Wi-Fi, including devices your neighbors add) can affect the way your Wi-Fi performs overall and/or in certain places in your home. You should test and adjust for these changes.

I recommend thinking about how your home network will evolve over the years. Will you add a lot of devices over the next few years? Will they be wired, wireless, or both? Are you going to want faster Internet or network speeds? Are there parts of your home network you'd like to upgrade? Technology is always advancing, and over time there will be better home networking gear that you'll want to take advantage of. You can step back through part 2 of this book with new goals in mind to plan improvements to your home network.

The Future of Home Networking

At the time of the writing of this book, a good solid home network includes at least 802.11ac wireless and gigabit Ethernet connections. Soon, 10 gigabit connections will be the standard, and Wi-Fi 6 and 6E will generate the speeds people are looking for. These are technologies you should start looking at and planning for, because they are already here.

Another technology that will change home networking is 5G. This newer cellular network standard, which is already available in limited forms and markets, has the potential to be as fast and as reliable as a home wired Internet connection. Some believe that 5G will replace your home Wi-Fi network, but I believe it's more likely the two will continue to work in tandem.

IPv6 (Internet Protocol version 6) is the replacement for the IPv4 standard that most networks use. The world is going to run out of IPv4 addresses, and IPv6 provides billions more IP addresses. It also can change how networking works. For instance, if you convert your home network to IPv6, you'll no longer need to configure NAT. Most ISPs and network gear already support IPv6. IPv4 in home networks will continue working for the foreseeable future, but IPv6 is still the future of IP.

The demands on the home network will continue to grow over time. Streaming and cloud services that require good home networks and Internet service will continue to increase. What seems like plenty of bandwidth today inevitably will seem constrained in the future.

The biggest area of future home networking that you should concern yourself with is network security. Hacking networks for sensitive information that can facilitate identity theft and ransomware have become increasingly lucrative over the last few years.

The good news about the future of home networking is that wired and wireless connections will become faster and more stable, support more simultaneous devices, and be easier to set up. You'll rely more on your home network than ever, but it will be up to it. The bad news is that you'll have to stay more vigilant about home network security than ever, as more and more services depend on it, and more and more bad actors gain the tools and skills to try and compromise your network. You'll be prepared for this future if you follow the steps and tools provided in this book.

Key Takeaways

❏ It takes maintenance and troubleshooting to keep your home network running smoothly over time.

❏ Common issues you may need to troubleshoot include the following:

　❏ A computer on your network can't access shared files and printers.

　❏ A computer on your network can't access the Internet.

　❏ Your computer's network speeds are slow.

❏ Common Wi-Fi issues you may need to address include the following:

　❏ Diagnosing and fixing slow Wi-Fi speeds

　❏ Fixing common Wi-Fi security problems

❏ You can get help fixing your home network from the following places:

    ❏ Home network device manufacturers

    ❏ Home network device support forums

    ❏ Large support organizations

    ❏ Local home network help companies

    ❏ Social media

    ❏ Home network help blogs and websites

❏ Keeping the devices on your home network up to date is important.

❏ Home network technology will continue to evolve and improve rapidly.

APPENDIX A – SAMPLE HOME NETWORK BUILDOUTS

Below are sample devices I recommend you consider depending on the size of your home. I don't list specific models because they change too quickly. However, you can always find my recommended models and reviews at the HomeTechHacker Shop[5] and in the HomeTechHacker Buyer's Guide[6].

Apartment/Small Home Build

Apartments and smaller homes (less than 2,000 square feet) generally don't need a lot of equipment for a good home network. The key issues to solve with apartments, and smaller homes to a lesser extent, is Wi-Fi interference from neighbors' network equipment and other devices. For this reason, I recommend a Wi-Fi 6 or 6E router, as those protocols are significantly less subject to interference. Just remember, you'll need client devices that can take advantage

---

[5] https://hometechhacker.com/shop/
[6] https://hometechhacker.com/top-picks-for-common-household-tech/

of the newer protocols. However, the routers will work with older devices too.

Device	Purpose/Capability	Suggested Brands
Wi-Fi 6 or 6E router	To distribute Internet access (wired and wireless) throughout your home, provide a shared network for your devices, and a first line of defense (firewall) for your home network.	Ubiquiti TP-Link Asus Netgear Linksys
Gigabit network switch	Provide additional wired connections to your home network.	TP-Link Netgear Trendnet D-Link
Cat 6 wiring	Make wired connections to your switch and router from your computers.	AmazonBasics Cable Matters Mediabridge DBallionDa

Mid-size Home Build

The needs of mid-sized homes (2,000–3,000 square feet) are often greater than those of small homes. The space that the Wi-Fi signal needs to cover pushes the boundaries of what most Wi-Fi routers are capable of. In these cases, an access point or a mesh Wi-Fi router may be needed to make sure there is good Wi-Fi everywhere. Make sure your Wi-Fi router supports at least the Wi-Fi 5 (Wireless AC) protocol.

If you are running Ethernet cables in the home, choose Cat 6A or better. The Cat 6A wiring will help prepare you for faster connections (e.g., 10 gig). Cat 6 and Cat 5e can support the faster speeds but not for as long of a distance. If the house is larger, some of your network cable runs may need Cat 6A.

Device	Purpose/Capability	Suggested Brands
Wi-Fi router (consider a mesh router)	To distribute Internet access (wired and wireless) throughout your home, provide a shared network for your devices, and a first line of defense (firewall) for your home network.	Ubiquiti TP-Link Asus Netgear Linksys Google Eero
Wi-Fi access point	To distribute your wireless signal in your home.	Ubiquiti TP-Link Netgear Linksys
Gigabit network switches	Provide additional wired connections to your home network.	TP-Link Netgear Trendnet D-Link
Cat 6A wiring or better	Make wired connections to your switch and router from your computers.	AmazonBasics Cable Matters Mediabridge DBallionDa

Large Home Build

Larger homes (greater than 3,000 square feet) require a similar level of equipment as mid-sized homes do, just more of it. However, if you are outfitting the home with network equipment for the first time, I recommend preparing for 10 gig networks. Rebuilding the network infrastructure of a house so large could take a lot of time and energy. Focus on putting in good wiring and capable switches in hard-to-access places. Prices will steadily go down in the future; however, 10 gigabit is significantly more costly now.

Also, owners of larger homes often want surveillance systems. A PoE switch that provides power to outdoor locations can make installing a good networked surveillance system a lot easier.

Device	Purpose/Capability	Suggested Brands
Wi-Fi router (consider a mesh router)	To distribute Internet access (wired and wireless) throughout your home, provide a shared network for your devices, and a first line of defense (firewall) for your home network.	Ubiquiti TP-Link Asus Netgear Linksys Google Eero

Wi-Fi access points	To distribute your wireless signal in your home.	Ubiquiti TP-Link Netgear Linksys
Gigabit (or better) network switches (consider a POE switch)	Provide additional wired connections to your home network.	TP-Link Netgear Trendnet D-Link MikroTik
Cat 6A wiring or better	Make wired connections to your switch and router from your computers.	AmazonBasics Cable Matters Mediabridge DBallionDa

APPENDIX B – CHECKLISTS

Wi-Fi Router Checklist

Wi-Fi router features to look for:

❏ Wireless AC support

❏ MU-MIMO or SU-MIMO support

❏ Advanced firewall

❏ Quality of Service (QoS)

❏ Support of at least WPA2 AES security

❏ Gigabit Ethernet ports

❏ Guest network support

❏ Receipt of consistent firmware updates

❏ *Not* an ISP-provided model

❏ Well reviewed on Amazon or in professional online reviews

Home Network Security Checklist

❏ Your router firewall is turned on.

❏ Your router is less than five years old or still gets frequent updates.

❏ Your router firmware is up to date.

❏ You are using WPA2 or WPA3 security with AES.

❏ You have a strong Wi-Fi password.

❏ You've changed your Wi-Fi password in the last year.

❏ You have a guest network.

❏ You've isolated your Internet of Things (IoT) devices.

❏ Your router web interface isn't accessible from the Internet.

❏ Your Wi-Fi network isn't set to the default name.

❏ You know all the devices on your home network, and monitor your home network for new devices.

❏ Your antivirus/malware software is up to date.

❏ You've encrypted the hard drives of your mobile devices (phones, tablets, laptops, etc.).

❏ You've centralized your computer logs, and monitor them for security and failure events.

❏ You regularly back up important files.

❏ You have a written disaster recovery plan for your home network in case of emergency.

Home Router Configuration Checklist

❏ Determine the optimal location for your Wi-Fi signal.

❏ Connect the router to your Internet connection.

❏ Change the router's default username and password.

❏ Check for firmware updates.

❏ Configure DHCP.

❏ Configure the router's Wi-Fi networks.

❏ Configure the router's security (turn on the firewall).

❏ Disable remote administration access.

❏ Configure QoS.

❏ Test your Internet speed.

Home Network Evaluation Checklist

❑ How fast is your Internet service? Is it fast enough?

❑ Is your Internet service reliable?

❑ Is your current networking hardware reliable and fast enough?

❑ Is your Wi-Fi router still receiving security updates?

❑ Does your Wi-Fi frustrate you?

❑ Do you know all the devices on your network?

❑ Is your network strong enough for as many devices as you have?

New Construction Home Network Checklist

❏ Create at least one wired connection with high-quality cable (Cat 6A or better) to every room.

❏ Look for places to install wireless access points in the ceiling and on a wall and run network cable to those positions.

❏ Consider Ethernet runs to locations where you may want to install video surveillance cameras, including outdoors.

❏ Install a PoE-capable switch in the location where your Ethernet will terminate.

❏ Make sure potential ISPs make connections or create an easy way to make connections to your home.

APPENDIX C – NETWORK DIAGRAMS

Below is a basic network diagram including a modem, a router, a few wireless devices, and a network switch for additional wired devices. This is a common configuration.

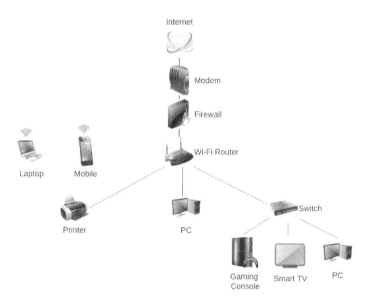

Basic network diagram

Below is an advanced network diagram, which expands on the basic network by including an additional switch, an access point, and additional devices. This is a common configuration for larger home networks. Some may even include additional switches and access points.

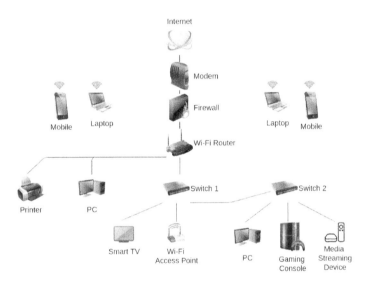

Advanced network diagram

Below is a network diagram for a mesh Wi-Fi system. This diagram reflects a system utilizing a mesh Wi-Fi router and satellites to provide a strong wireless connection to all devices throughout the home.

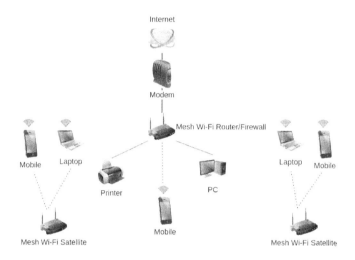

Mesh Wi-Fi network diagram

Below is a diagram for a home network using powerline
adapters. One powerline adapter connects to a network
switch that is wired to the router. The other powerline
adapter connects to the network via the home electrical wiring. A network switch connected to the powerline adapter
provides wired networking to the connected devices.

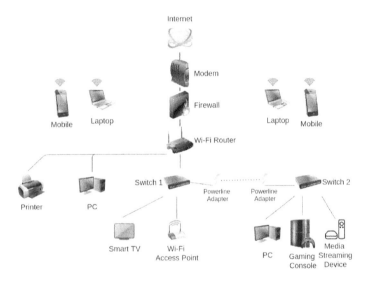

Network diagram for powerline adapters

APPENDIX D – ADDITIONAL RESOURCES

- **HomeTechHacker Blog**
  (https://hometechhacker.com/) – This is my personal blog where you can find many helpful articles about building and improving your home network and smart home.
- **HomeTechHacker Shop**
  (https://hometechhacker.com/shop/) – Here, I maintain an up-to-date list of recommended home network and smart home devices.
- **CNET Home Internet**
  (https://www.cnet.com/home/internet/) – This site has an abundance of information and reviews of home network products.
- **The Ambient**
  (https://www.the-ambient.com/) – This site has lots of up-to-date reviews on the latest home network products.
- **Digital Trends**
  (https://www.digitaltrends.com/smart-home-reviews/) – Their home technology product reviews are thorough and entertaining.

- **pfSense Documentation**
  (https://docs.netgate.com/pfsense/en/latest/) – I've been really happy with pfSense software running my router. It's free software with lots of documentation and features that can run on hardware of your choice. You can buy a router with pfSense pre-installed. This documentation will get you started.

- **SmallNetBuilder**
  (https://www.smallnetbuilder.com/) – SmallNet-Builder is a great site for those into making their home networks the best they can be. It is filled with tutorials, reviews, and an extremely helpful forum for technical assistance.

- **ZDNet**
  (https://www.zdnet.com/) – They have tons of product reviews about everything home technology, including home network products.

- **Subnet Mask Cheat Sheet**
  (https://www.pcwdld.com/subnet-mask-cheat-sheet-guide) – This is a thorough explanation and guide for understanding IP address schemes and subnets.

- **How to Create a Strong Password (and Remember it)**
  (https://www.howtogeek.com/195430/how-to-create-a-strong-password-and-remember-it/) – This is a great resource for creating good passwords from How-To Geek.

GLOSSARY

AES – AES (i.e., Advanced Encryption Standard) is a tech-nology used to encrypt data and secure transmission of data between devices. Wi-Fi networks use AES to secure data transmission.

Band Steering – Band steering is a technology used by some Wi-Fi equipment that encourages client devices to use 5 GHz networks instead of 2.4 GHz networks if the device supports both networks. This helps relieve congestion on 2.4 GHz networks and improves the overall Wi-Fi quality for all devices.

Cat 5e, Cat 6, Cat 6a, Cat 7, Cat 8 – These are all different standards of Ethernet cable, which are the backbone of the wired network in your home. The higher the number, the later the Ethernet standard. The later the Ethernet standard, the faster speeds the cable will support over longer dis-tances. If you are laying down new wire at the time of my writing this book, try to get at least Cat 6a, which can sup-port a 10-Gigabit connection for 100 meters.

DHCP – DHCP (i.e., Dynamic Host Control Protocol) is a network management protocol used to automate configuring device IP addresses. This allows each device to communicate with other devices easily. Devices can have IP addresses that aren't assigned by a DHCP server (static IP addresses), but these must be configured manually. As networks get larger, it's easier to manage IP addresses, using a DHCP server. Usually, your router functions as the DHCP server on your network.

Firewall – A firewall is a network security device that monitors incoming and outgoing network traffic and decides which parts of that traffic to block from entering or exiting your network, based on a set of rules. It is primarily used to protect your network from threats coming from the Internet. A firewall can be a standalone device, but it is most often a function available in your router.

Gateway – A gateway is a network device that connects two networks with different transmission protocols together. In a home network setting, your gateway is most likely the modem or modem/router combo your ISP provides.

IoT – IoT (i.e., Internet of Things) is a collection of interconnected and interrelated devices that can communicate and transfer data over various networks without human interaction. It most commonly refers to any device that connects to the Internet. Examples include smart TVs, smart speakers, toys, wearables, smart appliances, and smart meters.

IP Address – An IP address (i.e., Internet Protocol address) is a numerically based label that uniquely identifies a device on a network.

ISP – ISP (i.e., Internet Service Provider) is a company that provides Internet as a service to a home or business. These are often cable and other telecommunications companies like Comcast, CenturyLink, and Verizon.

LAN – LAN (i.e., local area network) is a computer network that usually connects computers in a specific geographic location—a home, a school, or an office building.

MAC Address – A MAC address (i.e., a media access control address) is a hexadecimal address assigned to network devices. MAC addresses are typically unique to a particular device and can be used to identify devices on a network and to assign devices unique IP addresses.

Mesh Wi-Fi Router – A mesh Wi-Fi router, sometimes called whole home mesh Wi-Fi, consists of a wireless router and satellite wireless access points that provide wireless access throughout your home. The satellite wireless access points connect to the router wirelessly and don't need a wired network connection to the router. This means they can be flexibly placed at optimal points to provide Wi-Fi throughout a home.

Modem – Modem, which is short for modulator-demodulator, converts signals from one type of device to another type of device. One example is a cable modem, which converts signals from a coaxial cable (analog) to an Ethernet signal (digital) that routers typically use.

NAT – NAT (i.e., network address translation) allows a single device to act as an agent between the Internet and your local network. It enables private IP networks (like most home networks) to share a public IP address and Internet connection. A router usually implements NAT, allowing all of the computers on your network to have a private IP address but appear as a single address when accessing Internet resources such as websites.

NIC – A NIC (i.e., network interface card) is a device that connects a computer to a computer network. In a home network, a NIC is the device in a computer that allows it to physically or wirelessly connect to your LAN.

QoS – QoS (i.e., quality of service) is a technology that manages data traffic in a way that allows different types (e.g., video streaming, video conferencing, online gaming, web browsing, etc.) to be appropriately prioritized so that each type of traffic works well when the other types of traffic are simultaneously on the network.

Router – A router forwards (or routes) packets of data between different networks. For instance, the router in your home routes packets coming from your ISP (WAN) to your private home network (LAN) and vice versa.

Smart Home – A smart home is a home that provides some combination of comfort, energy efficiency, security, lighting, etc., aided by technology that allows these systems to be automated, integrated, and available for remote control.

Switch – A switch, or a network switch, is a device in a computer network, which connects multiple devices (e.g., computers, access points, printers, etc.) together.

VPN – A VPN (i.e., virtual private network) allows a device to communicate securely across a public network with a private network (such as your home network). VPNs work by encrypting the data that travels over the Internet between a device and the private network. VPNs can also be used to connect private networks together securely over the Internet.

WAN – WAN (i.e., wide area network) is a computer network that covers a large geographic region. WANs are similar to LANs but aren't limited to a single location and are usually larger.

Wi-Fi – Wi-Fi is the name of a wireless networking technology that uses radio waves to provide high-speed network connections.

Wi-Fi Extender – A Wi-Fi extender, sometimes called a Wi-Fi repeater, extends the range of a Wi-Fi network by connecting to your Wi-Fi router and repeating the Wi-Fi signal to another area.

Wireless Access Point – A wireless access point connects to a router, switch, or hub via Ethernet, and emits a Wi-Fi signal for connecting to your network. Wireless access points are used for extending a wireless network.

WLAN – WLAN (i.e., wireless local area network) is just a wireless version of a LAN, which is a computer network that usually connects computers in a specific geographic location—a home, a school, or an office building.

WPA – WPA (i.e., Wi-Fi Protected Access) is a security standard used to protect Wi-Fi network access. The original WPA standard is considered insecure. All your devices should be using WPA2 or WPA3.

ABOUT THE AUTHOR

Marlon Buchanan has worked in the IT field for over twenty-five years as a software developer, a college instructor, and an IT Director. In his free time, he can be found researching new smart home and home network projects, watching and playing sports with his children, and writing articles for his blog, HomeTechHacker.com. He holds a bachelor's degree in computer engineering and master's degrees in software engineering and business administration.

Please sign up for his newsletter on his blog. You can also follow him on these social media channels:

- Twitter – Twitter.com/HomeTechHacker (@HomeTechHacker)
- Pinterest – Pinterest.com/HomeTechHacker
- Facebook – Facebook.com/HomeTechHacker

OTHER BOOKS BY MARLON BUCHANAN

These books are available in various stores worldwide, including on Amazon.com:

- *The Smart Home Manual: How to Automate Your Home to Keep Your Family Entertained, Comfortable, and Safe*

- *Home Wi-Fi Tuneup: Practical Steps You Can Take to Speed Up, Stabilize, and Secure Your Home Wi-Fi*

WHAT DID YOU THINK OF THE HOME NETWORK MANUAL?

First of all, thank you for purchasing *The Home Network Manual*. I know that you could have picked any number of books to read, but you picked this book, and for that I am extremely appreciative.

I hope that it has inspired you and helped build and improve your home network. If so, it would be really nice if you could share this book with your friends and family by posting to Twitter, Facebook, and Pinterest.

If you enjoyed this book and found some benefit to reading this, I'd love to hear from you. I hope that you can take the time to post a review on Amazon. Your feedback and support will help me greatly improve my writing craft for future projects.

I want you, the reader, to know that your review is very important and very appreciated.

I wish you all the best in your future home network success!

Made in the USA
Monee, IL
19 January 2023